BIOCHEMICAL SOCIETY SYMPOSIA

No. 39

CALCIUM AND CELL REGULATION

*BIOCHEMICAL SOCIETY SYMPOSIUM No. 39
held at the University of Birmingham, April 1973*

Calcium and Cell Regulation

ORGANIZED AND EDITED
BY
R. M. S. SMELLIE

1974

LONDON: THE BIOCHEMICAL SOCIETY

The Biochemical Society,
7 Warwick Court,
London WC1R 5DP, U.K.

Copyright © 1974 by The Biochemical Society: London

ISBN 0 9501972 6 2

All rights reserved
No part of this book may be reproduced in any form by photostat, microfilm
or any other means, without written permission from the publishers

Printed in Great Britain by William Clowes & Sons Limited,
London, Colchester and Beccles

List of Contributors

C. C. Ashley (*Department of Physiology, University of Bristol, Bristol BS8 1UG, U.K.*)

P. C. Caldwell (*Department of Zoology, University of Bristol, Bristol BS8 1UG, U.K.*)

E. Carafoli (*Institute of General Pathology, University of Modena, 41100 Modena, Italy*)

P. Cohen (*Department of Biochemistry, Medical Sciences Institute, University of Dundee, Dundee DD1 4HN, U.K.*)

R. M. Denton (*Department of Biochemistry, University of Bristol Medical School, Bristol BS8 1TD, U.K.*)

W. W. Douglas (*Department of Pharmacology, Yale University School of Medicine, New Haven, Conn. 06510, U.S.A.*)

N. M. Green (*National Institute for Medical Research, London NW7 1AA, U.K.*)

P. M. D. Hardwicke (*National Institute for Medical Research, London NW7 1AA, U.K.*)

H. T. Pask (*Department of Biochemistry, University of Bristol Medical School, Bristol BS8 1TD, U.K.*)

S. V. Perry (*Department of Biochemistry, University of Birmingham, Birmingham B15 2TT, U.K.*)

P. J. Randle (*Department of Biochemistry, University of Bristol Medical School, Bristol BS8 1TD, U.K.*)

D. L. Severson (*Department of Biochemistry, University of Bristol Medical School, Bristol BS8 1TD, U.K.*)

D. A. Thorley-Lawson (*National Institute for Medical Research, London NW7 1AA, U.K.*)

R. J. P. Williams (*Inorganic Chemistry Laboratory, University of Oxford, Oxford OX1 3QR, U.K.*)

Preface

This is the first occasion on which the Annual General Meeting of the Biochemical Society has been held in Birmingham and it is appropriate that the nature of this Symposium should have been suggested by members of the Department of Biochemistry of Birmingham University. The role of calcium in regulatory processes is also very relevant to the work of some research groups in Birmingham, and I am particularly grateful to Professor S. V. Perry and to Dr. J. B. Finean for their help in selecting topics and speakers.

The researches of many biochemists have come to be concerned with the control of cellular function and metabolism, and many different mechanisms have been discovered during the last decade. Some biochemists, including myself, are less familiar with control exerted through the movement of ions and variations in ionic concentration, and it is to be hoped that this Symposium on the behaviour to calcium ions may assist them to reach a clearer understanding of such mechanisms.

The Biochemical Society is grateful to Professor S. V. Perry and Dr. R. J. P. Williams for acting as Chairmen and to the distinguished group of contributors who deserve the main credit for making the Symposium a success.

R. M. S. SMELLIE

Institute of Biochemistry,
University of Glasgow,
Glasgow G12 8QQ,
U.K.

Contents

	Page
List of Contributors	v
Preface	vii

Involvement of Calcium in Exocytosis and the Exocytosis–Vesiculation Sequence
By W. W. Douglas .. 1

Calcium Movements in Relation to Contraction
By C. C. Ashley & P. C. Caldwell .. 29

The Role of Phosphorylase Kinase in the Nervous and Hormonal Control of Glycogenolysis in Muscle
By P. Cohen .. 51

Calcium Ions and the Regulation of Pyruvate Dehydrogenase
By P. J. Randle, R. M. Denton, H. T. Pask & D. L. Severson .. 75

Mitochondrial Uptake of Calcium Ions and the Regulation of Cell Function
By E. Carafoli .. 89

Preliminary Studies on the Structure of the Calcium-Activated Adenosine Triphosphatase of Sarcoplasmic Reticulum
By N. M. Green, P. M. D. Hardwicke & D. A. Thorley-Lawson .. 111

Calcium Ions and the Function of the Contractile Proteins of Muscle
By S. V. Perry .. 115

Calcium Ions: Their Ligands and their Functions
By R. J. P. Williams .. 133

Author Index .. 139

Subject Index .. 145

Involvement of Calcium in Exocytosis and the Exocytosis–Vesiculation Sequence

By WILLIAM W. DOUGLAS

Department of Pharmacology, Yale University School of Medicine, 333 Cedar Street, New Haven, Conn. 06510, U.S.A.

Synopsis

There is increasing evidence that calcium has a critical role as a mediator in stimulus–secretion coupling in a great variety of cells—nerves, neurosecretory fibres, endocrine, exocrine and others—where secretion involves extrusion of preformed product from membrane-limited granules or vesicles, and that calcium functions in each instance to activate exocytosis. The present paper reviews the status and general applicability of the 'calcium–exocytosis' hypothesis, introduces some variations on the theme, discusses the phenomenon of formation of microvesicles ('synaptic vesicles') as a consequence of exocytosis, and indicates some experimental approaches to the nature of calcium's involvement in exocytosis and resulting vesiculation under the following headings. 1. Development of the concept of exocytosis and its association with calcium. 2. Status of the 'calcium-influx exocytosis' hypothesis of stimulus–secretion coupling in chromaffin cells. (i) Calcium influx as the mediator of secretion. (ii) Exocytosis as the secretory mechanism activated by calcium. 3. General applicability of the concept of calcium-activated exocytosis. 4. Case for calcium-activated exocytosis in neurosecretory fibres and neurons. (i) Exocytosis and synaptic vesicle formation. (ii) Calcium influx necessary and sufficient. 5. Unity of events in stimulus–secretion coupling in neurons, neurosecretory fibres and chromaffin cells. 6. Variations on the calcium-activated exocytosis theme. (i) Secretion to excess of potassium and voltage-dependent calcium influx. (ii) Different sources of calcium: mobilization and influx. (iii) Serial fusion of granules: 'compound' versus 'simple' exocytosis. 7. Aspects of exocytosis illustrated by microcinematography of mast cells. 8. Possible involvement of calcium in 'vesiculation coupled to exocytosis'. 9. Calcium and 'contractile' hypotheses of secretion. (i) Granule ATPase (adenosine triphosphatase). (ii) Microtubules, microfilaments, colchicine and cytochalasin. 10. Concluding remarks.

Development of the Concept of Exocytosis and its Association with Calcium

Almost 20 years ago, Bennett (1956) speculated on theoretical grounds that if cells could perform pinocytosis, that is incorporate extracellular material by invagination and pinching off of surface membrane from membrane-limited droplets, then they could conceivably operate in the reverse manner and incorporate the membrane of cytoplasmic droplets into the cell surface. In this scheme of 'membrane flow' involving 'reverse pinocytosis' (see also De Robertis *et al.*, 1965) there is anticipation of the subject I wish to discuss today, the phenomenon known variously as 'reverse micropinocytosis', 'emiocytosis', or now most commonly, as 'exocytosis' a term devised by de Duve (1963). It was, of course, the introduction of electron microscopy allowing detection of cell membranes that permitted speculation and experiments along these lines and by the late 1950's and early 1960's there were indications that some secretory cells utilized this phenomenon of 'reverse micropinocytosis' to export secretory material. By this time it had become clear that in many secretory cells the product was sequestered in membrane-limited granules or vesicles of various sizes, and Palade (1958) had demonstrated electron-microscopic images from pancreatic acinar cells showing direct discharge of material from zymogen granule to the cell exterior by a process seemingly involving the fusion of granule membrane with the cell surface followed by rupture of the fused membranes. Moreover, there were indications that reverse micropinocytosis could also operate in endocrine cells, early examples being chromaffin cells of the adrenal medulla (De Robertis & Vaz Ferreira, 1957; De Robertis & Sabatini, 1960), pancreatic β-cells (Williamson et al., 1961), and adenohypophyseal cells (Farquhar, 1961). While the morphologists were building up this evidence that exocytosis was a mechanism of widespread occurrence clues were rapidly accumulating implicating calcium in the exocytotic process. The first set of clues came from a study Ronald Rubin and I had undertaken in an attempt to define how secretory stimulants (secretagogues) acted on cells (Douglas & Rubin, 1961). For this purpose we selected the chromaffin cells of the adrenal medulla, a favourite test object of pharmacologists and physiologists for many years, which offered a variety of technical advantages. Our analysis, about which more will be said below, had indicated that the secretory response was uniquely and quantitatively dependent on the concentration of calcium in the extracellular environment and that calcium ions themselves could elicit secretion under conditions where it seemed reasonable to suppose they could penetrate the chromaffin cell membranes. On these and other grounds, which will be mentioned later, we advanced the hypothesis that the secretagogue acetylcholine functions simply by some action on the plasma membrane increasing its permeability and allowing inward movement of calcium ions. In this calcium-influx hypothesis of secretion there was clearly a parallel with events in muscle and we suggested that calcium might act in what we called 'stimulus–secretion coupling' in a manner comparable with its mediator function in excitation–contraction coupling. Our reason for preferring 'stimulus' to 'excitation' in regard to secretion has been discussed elsewhere (Douglas, 1968a). At the time there was evidence that secretion in cells as diverse as neurons and mast cells was dependent on extracellular calcium and we proposed that calcium entry might be a general mechanism for releasing biologically active substances. The next question to be confronted was the nature of the process set in motion by the calcium ions penetrating the cell membrane. Although exocytosis was but one of many theories of catecholamine release that had then been advanced (see Douglas, 1966), for various reasons we favoured the idea that calcium was inducing some active process and exocytosis seemed an attractive possibility (Douglas & Rubin, 1961). In retrospect the evidence for exocytosis in medullary chromaffin cells at the time was certainly rather flimsy being based entirely on electron-microscopic images lacking the convincing qualities of those more recently obtained. In any event, in this work on the adrenal medulla there can be recognized the beginnings of what one might refer to as the concept of calcium-activated exocytosis. This concept was soon reinforced, however, when Alan Poisner and I extended the study of stimulus–secretion coupling to exocrine glands where the evidence for exocytosis was much better established. We chose the submaxillary salivary gland since it allowed us to test another secretagogue, noradrenaline, as well as acetylcholine. To our delight both secretagogues again proved calcium-dependent. The moment seemed ripe for suggesting that 'calcium might be generally involved wherever cells are extruding material by 'reverse micropinocytosis'; and as examples of where such an involvement might be sought we cited the β-cells of the endocrine pancreas and the cells of the anterior pituitary (Douglas & Poisner, 1963). Within a few years the anticipated calcium-dependence had been demonstrated in each of these endocrine glands (see review by Rubin, 1970). Meanwhile further analysis of stimulus–secretion coupling in the chromaffin cell had buttressed the calcium-influx hypothesis: calcium had been shown to be not merely an essential but a sufficient ionic requirement for catecholamine secretion (Douglas & Rubin, 1963). Accumulation of ^{45}Ca on exposure to acetylcholine had been observed (Douglas & Poisner, 1962). Intracellular microelectrode techniques had demonstrated unequivocally the membrane action of acetylcholine and yielded evidence of inward depolarizing current carried by calcium ions, and local anaesthetics blocking ^{45}Ca uptake (Rubin et al., 1967) or selectively blocking inward calcium current (Douglas & Kanno, 1967) inhibited secretion. This evidence is set out in greater detail below. During the same period exocytosis was established beyond all reasonable doubt as the main and probably sole mechanism of catecholamine secretion as a result of two complementary approaches. The first, introduced in 1965, was the principle of deducing the nature of the cellular events involved in catecholamine secretion by analysing the venous effluent from perfused adrenal glands for various cell constituents other than the catecholamines (see Douglas, 1966). The initial results showing escape of ATP metabolites *pari passu* with the catecholamines (Douglas et al., 1965; see also Douglas & Poisner, 1966a,b) were soon followed by the discovery that the effluent also contained the protein, chromogranin (Banks & Helle, 1965). Clearly the ATP-rich, chromogranin-containing chromaffin granules (and not extragranular 'pools') were directly involved in secretion, moreover, since neither class of compound should pass through intact membranes the granules apparently voided directly to the extracellular environment. By 1967 further experiments in several laboratories had rounded off the evidence and made it clear there was preferential release of soluble granule contents without discharge of the insoluble constituents associated with the granule membranes; and further, that all this took place without evidence of increased permeability of the cell membrane to small proteins in the cytoplasm. The only mechanism which had been described in any secretory cell that would account for such results was exocytosis (see review by Douglas, 1968a). The second approach, electron microscopy, clinched the matter. Early work had provided some images suggestive of exocytosis in chromaffin cells of several species and when Diner (1967) examined hamster adrenals she obtained unequivocal and abundant evidence of the process.

By 1967 then, it was possible to conclude with some confidence that calcium's role in the medullary chromaffin cell was indeed concerned with exocytosis. Moreover, it was by this time apparent that calcium was required for the secretory activity of a wide spectrum of cells where the only common factor seemed to be storage of secretory product in membrane-limited granules or vesicles. This led to the specific proposal in a lecture entitled 'Stimulus–Secretion Coupling: The Concept and Clues from Chromaffin and Other Cells' (Douglas, 1968a), that the basic pattern discernible in the chromaffin cell might operate in the others, that the common requirement for calcium was most simply explained by its having a common mediator role and setting in motion a common secretory mechanism, exocytosis. This, it was argued, was 'plausible since exocytosis in each cell involves only two components, the membrane delineating the cell, the plasmalemma, and the membrane surrounding the secretory granule, and there is no impediment to supposing that these two components are basically similar in different cells. On this view then the chemical nature of the content of the granules which varies from one cell to the next is irrelevant to the secretory process; and calcium's function is concerned solely with the events governing delivery of the contents to the cell exterior. The calcium-dependent process could then be likened to the postman who delivers packages without regard (hopefully!) to their contents.' The range of cells where it was proposed that such a mechanism might operate included neurons, neurosecretory fibres, endocrine, and exocrine cells and 'odd' secretory cells such as the mast cell. And the stimuli included action potentials, chemical transmitters, hormones, autacoids of various sorts, bacterial toxins and antigens. Further, given the many obvious parallels between stimulus–secretion coupling and excitation–contraction coupling it seemed 'at least conceivable that nature uses similar devices at the molecular level to effect these two similarly disparate responses, contraction and secretion'.

In the next 5 years results obtained from studies on a great variety of cells with different techniques greatly strengthened this general concept of calcium-activated exocytosis and demonstrated the heuristic value of the chromaffin cell model, although some variations on the theme have become apparent. With increasing evidence of the operation of exocytosis, attention has been drawn to the problem of how the cell disposes of the 'excess' granule membrane incorporated into the cell surface. And through all of this there has run the challenge of defining at fine-structural or molecular levels the precise nature of calcium's critical function.

With increasing confidence that the chromaffin cell is, as early suspected, a most valuable model I believe it appropriate to begin by reviewing the uniquely extensive evidence this cell has provided.

Status of the 'Calcium Influx–Exocytosis' Hypothesis of Stimulus–Secretion Coupling in Chromaffin Cells

Calcium influx as the mediator of secretion

The observations supporting the calcium-influx hypothesis of stimulus–secretion coupling in the chromaffin cells (Douglas & Rubin, 1961) have for the most part been discussed in review articles (Douglas, 1968a, 1974a; Rubin, 1970).

Simple enumeration here will suffice to indicate the extent and strength of the evidence.

1. Secretion in response to the physiological secretagogue acetylcholine fails rapidly when calcium is omitted from the extracellular environment (without need for chelating agent) and recovers promptly when calcium is restored (Douglas & Rubin, 1961).

2. A similar calcium requirement can be shown for a variety of nicotinic and muscarinic drugs related to acetylcholine and various other unrelated medullary secretagogues, amines and polypeptides (Poisner & Douglas, 1966).

3. Intracellular recording shows that acetylcholine and examples from each of these other classes of secretagogues act on the chromaffin cell membrane as judged by their depolarizing effects (Douglas et al., 1967a).

4. Inward depolarizing current in response to acetylcholine increases with increasing $[Ca^{2+}]_o$ and $[Na^+]_o$ (where 0 indicates extracellular concentrations) over a wide range and is thus carried by both ions (Douglas et al., 1967b).

5. As $[Ca^{2+}]_o$ is increased there is a corresponding rise in the secretory response to acetylcholine. By contrast a decrease of $[Na^+]_o$ (and acetylcholine-evoked Na^+ influx) potentiates acetylcholine-evoked catecholamine output (Douglas & Rubin, 1963), and has been shown, in other cells, to favour calcium influx and elevated $[Ca^{2+}]_i$ (where i indicates intracellular concentrations) (references in Baker, 1972).

6. Mg^{2+} and Co^{2+}, ions known to decrease calcium influx in invertebrate axons (references in Baker, 1972), inhibit secretion to acetylcholine (Douglas & Rubin, 1963, 1964a). And interaction between effects of calcium and magnesium on the chromaffin cell has been shown directly (see point 10 below).

7. Local anaesthetic in a concentration allowing inward Na^+ current but blocking inward Ca^{2+} current (Douglas & Kanno, 1967) [or ^{45}Ca uptake (Rubin et al., 1967)] also inhibits secretion.

8. Secretory responses to acetylcholine can be elicited when the external medium consists simply of sucrose and Ca^{2+} (with Cl^- or the impermeant anion SO_4^{2-}); and this effect is lost when Ca^{2+} is removed from the medium and inward Ca^{2+} current thereby abolished (Douglas & Rubin, 1963; Douglas et al., 1967b).

9. Elevation of $[K^+]_o$, which is known to depolarize various cell membranes and increase permeability to Ca^{2+} also depolarizes chromaffin cells (Douglas et al., 1967b), increases their ^{45}Ca uptake (Douglas & Poisner, 1961), and elicits a secretory response resembling that to acetylcholine in its dependence on $[Ca^{2+}]_o$ (Douglas & Rubin, 1961), requirement for no other ion, and inhibition by Mg^{2+} (Douglas & Rubin, 1963) and local anaesthetic (Rubin et al., 1967).

10. Calcium itself in the normal physiological concentration (2mM) elicits a massive secretory response when introduced to chromaffin cells whose membrane permeability has been increased by a period of calcium deprivation (Douglas & Rubin, 1961). Calcium is thus a secretagogue in conditions where it can seemingly penetrate. This response is blocked by magnesium (Douglas & Rubin, 1963).

11. Profound decrease in $[Na^+]_o$ or $[K^+]_o$ (Douglas & Rubin, 1961, 1963; Banks, 1970; Banks et al., 1969) or addition of ouabain (Banks, 1967) all increase basal as well as acetylcholine-evoked secretion and are all manoeuvres shown to pro-

mote intracellular accumulation of calcium in other tissues (references in Baker, 1972).

12. Metabolic inhibitors interfering with intracellular storage of calcium and which might therefore be expected to raise $[Ca^{2+}]_i$ have been observed to stimulate medullary secretion (Poisner & Hava, 1970).

13. Barium, an alkaline earth ion related closely to calcium, will itself induce secretion by some direct action on the chromaffin cell which requires the presence of no 'adjuvant' secretagogue (Douglas & Rubin, 1964a,b).

Although the above evidence (and other points made in the reviews cited) should be considered as a whole, the essence of it is simply that the physiological stimulus or secretagogue acetylcholine causes influx of calcium across the cell membrane, that calcium is a necessary and sufficient ionic condition for stimulus-evoked secretion, and that calcium itself can elicit secretion provided membrane permeability is increased.

Exocytosis as the secretory mechanism activated by calcium

The chemical evidence consistent with exocytosis was, as earlier noted, already impressive in 1967. Since then supplementary experiments and more refined techniques have completed the balance sheet and left little doubt that secretion involves discharge of the entire water-soluble content of the chromaffin granules with retention of the lipid granule membrane and its associated enzymes and other components and have established that this occurs without any escape of cytoplasmic marker molecules such as lactate dehydrogenase (see reviews by Kirshner & Kirshner, 1971; Kirshner & Viveros, 1970). Although this pattern is most neatly accounted for by exocytosis it is not, as some believe, proof of exocytosis. Alternative hypothetical schemes such as 'facilitated diffusion by membrane fusion' (Douglas, 1966) can be postulated (see Fig. 1 in Douglas, 1968a). Such a possibility is by no means far fetched since large molecules are known to traverse tight junctions between cells (Lowenstein, 1973). The chemical evidence must then be complemented by morphological evidence. Exocytosis is, after all, an essentially morphological concept. Diner's (1967) observation of abundant examples of exocytosis in chromaffin cells of the golden hamster is crucial and has been repeatedly confirmed (Benedeczky & Smith, 1972; Nagasawa & Douglas, 1972). It would be still more satisfactory if a correlation between stimulus and incidence of exocytosis could be shown in such material and Dr. Nagasawa and I have recently succeeded in this. In work on the exocytosis–vesiculation sequence to which I shall return later (Douglas & Nagasawa, 1971, 1972; Nagasawa & Douglas, 1972) we observed that excess of potassium, a convenient medullary secretagogue, facilitated our study by increasing the incidence of exocytosis. By comparing the number of exocytotic pits per unit length of cell surface in our electron micrographs we have found that there is a severalfold increase when glands incubated in 60mM-potassium–Locke's solution for several min are compared with control pieces of the same glands incubated in Locke's solution with conventional potassium concentration (Fig. 1). Moreover the number of exocytotic images produced by potassium increased as the extracellular calcium concentration was raised to 10mM or more. And, in a study employing freeze etching, Smith *et al.* (1973) report fewer exocytotic images in calcium-free conditions.

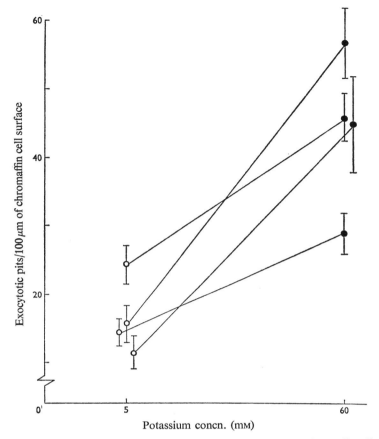

Fig. 1. *Stimulant effect of excess of potassium on exocytosis in chromaffin cells*

Hamsters were decapitated and one adrenal gland was removed and halved. One half was incubated for 2.5 min at 37°C in Locke solution, the other for the same time in high-potassium–Locke solution (60 mM-potassium with sodium decreased to maintain osmotic pressure). Electron micrographs were prepared (Nagasawa & Douglas, 1972) and random fields scanned for exocytotic pits along the vascular poles of the chromaffin cells. The results of four such experiments show that potassium increased the incidence of exocytosis (J. Nagasawa & W. W. Douglas, unpublished work).

Taken together then the chemical and morphological evidence now makes an almost unassailable argument that exocytosis is the main and probably only significant mechanism of catecholamine extrusion.

Summary

For the chromaffin cell then, the calcium influx–exocytosis hypothesis of stimulus–secretion coupling rests on a broad and firm base.

General Applicability of the Concept of Calcium-Activated Exocytosis

The question now to be considered is whether there is validity to the broader thesis that calcium-activated exocytosis operates in a wide spectrum of cells.

When this view was articulated more than 5 years ago the number of cells known or suspected to utilize exocytosis and to show calcium-dependence was rather limited although the diversity of cell types was already impressive (Douglas, 1968a). But since then the list has been greatly extended (see reviews by Rubin, 1970; Smith, 1971b) to embrace fresh examples from each of the categories discussed above: endocrine and exocrine systems, neurons, neurosecretory fibres and a variety of other cell types. None of the cells has been subjected to such intensive inquiry as the chromaffin cell—and here it should be recognized that the chromaffin cell is a particularly favourable object for study—nevertheless there is growing confidence that exocytosis is the principal, if not necessarily exclusive, mechanism of secretion in cells exporting membrane-packaged secretory product, and that calcium is a universal requirement, most authors subscribing to the view that calcium's role is intimately concerned with the activation of exocytosis. Rather than devote time to documenting this area I shall refer those interested to the several excellent reviews (Geschwind, 1971; Rubin, 1970; Smith, 1971b) and pursue instead the more controversial question of involvement of an essentially similar secretory mechanism in neurons and neurosecretory fibres. Although these are amongst the earliest cell types recognized to require calcium for their function there continues to be lively debate on whether they secrete by exocytosis (see Hubbard, 1970).

Case for Calcium-Activated Exocytosis in Neurosecretory Fibres and Neurons

It is somewhat paradoxical that the operation of exocytosis in neurosecretory fibres and neurons should be so controversial and so long refractory to experimental test. For these cells might be expected, more than any others, to conform to the chromaffin cell model because of their common developmental origins with chromaffin cells and their remarkably similar pattern of behaviour towards a spectrum of ionic manipulations and pharmacological agents. In arguing that these cells probably fell into the general pattern of calcium-activated exocytosis I stated some 6 years ago that 'when one compares the effect of the commonly occurring ions on neurones, chromaffin cells and neurosecretory cells it is difficult to avoid the conclusion these three developmentally related cells have a common mechanism for the release of their different secretory products' and further, 'it would be helpful to have the kind of chemical evidence that has clarified events in medullary cells' (Douglas, 1968a). More specifically I reiterated the proposal (Douglas, 1967) that if the protein neurophysin could be shown to escape from the neurohypophysis along with the posterior pituitary octapeptide hormones this would be presumptive evidence for the operation of exocytosis, despite the repeated failure to observe exocytosis in electron micrographs of neurohypophysial terminals.

Exocytosis and synaptic vesicle formation

When Fawcett *et al.* (1968) reported the release of neurophysin from stimulated dogs' pituitary glands my colleagues and I were persuaded that the failure to observe images of exocytosis by electron-microscopical methods must surely reflect some technical difficulty in capturing the process. Personal experience with the adrenal medulla strongly reinforced this suspicion: over a period of years beginning in 1963 with Professor René Couteaux and later in three other separate efforts with Dr. J. Rosenbluth, Dr. J. Osinchak and Dr. S. Malamed, I had sought electron-microscopical evidence of exocytosis in stimulated medullary cells in various species with a conspicuous (but not unique) lack of success. Yet when Dr. Nagasawa and I employed the golden hamster, the species first used successfully by Madame Diner, images of exocytosis were so plentiful they could not be missed. Since the chemical evidence for exocytosis had been provided by other species relatively refractory to demonstration of exocytotic images by electron microscopy the conclusion seemed obvious: there must be some species difference allowing exocytosis to be captured with relative ease in the golden hamster. Species difference could explain why exocytosis had been seen in neurosecretory systems of some invertebrates (Normann, 1965; Smith & Smith, 1966; Weitzman, 1969) but not in others (Scharrer & Weitzman, 1970). With this background, my colleagues and I undertook yet another electron-microscopic examination of posterior pituitary glands this time including material from golden hamsters. To our delight images of exocytosis, although sparse, were obviously present; and we succeeded in assembling a series of electron micrographs conforming to the classical descriptions of exocytosis in other cells. Moreover, some exocytotic images could also be detected in rats' neurohypophyses (Nagasawa et al., 1970; Douglas et al., 1971a). Others have since corroborated these findings both with conventional electron-microscopic methods (Krisch et al., 1972) and freeze etching (Santolaya et al., 1972). Although confirming that exocytosis can be detected, Krisch et al. (1972) do not agree with us in our interpretation. They hold that the paucity of images of exocytosis in neurohypophyseal terminals argues against this being an important mechanism of release of posterior pituitary hormones. But this argument carries little weight: it will be recalled that exocytotic images are rare in chromaffin cells of most species examined yet the chemical evidence favouring exocytosis is strong, and that only one species, the golden hamster, has yielded numerous examples of exocytosis. There is, moreover, no reason *a priori* why conventional fixation should be able to capture successfully such undoubtedly ephemeral events as exocytosis. A quantitative assessment of the importance of exocytosis in neurohypophysial terminals can scarcely be arrived at by counting the number of images of the process in electron micrographs. Some other criterion is needed. One such is the chemical evidence. Following the initial report by Fawcett et al. (1968) escape of neurophysin has been corroborated and studied in greater detail by a series of investigators and it is now clear that this protein escapes *pari passu* with the octapeptide hormones (Burton et al., 1971; Cheng & Freisen, 1970; Matthews et al., 1973; Nordmann et al., 1971; Uttenthal et al., 1971). Indeed discharge of neurophysin can be detected readily by immunological methods during physiological release of posterior pituitary hormones in humans (Legros et al., 1969). Admittedly, more chemical evidence will be required to establish exocytosis as the unique mechanism for release of posterior pituitary hormones—the chemical balance sheet does not yet rival in completeness that drawn up for the adrenal chromaffin cell—but obviously an impressive case can already be made. To this I feel one can add some new morphological evidence, which although indirect, fits nicely with the exocytosis concept.

What I intend to argue is that although exocytosis is difficult to detect by electron-microscopic methods in neurosecretory fibres the occurrence of the phenomenon is signalled by certain morphological consequences, its aftermath if you will. The argument is simply that just as condensation of water droplets in a cloud chamber allows 'visualization' of the passage of ionizing particles, so the appearance of certain entities in the neurohypophysial fibres are telltales of prior exocytotic activity. The objects to which I refer are the microvesicles (classically referred to as 'synaptic vesicles', Palay, 1957) whose function has so long been debated but which are known to increase in number after stimulation and hormone release. Our view is that these vesicles arise as a consequence of exocytosis by a micropinocytosis-like activity involving the exocytotic pit (Nagasawa et al., 1970; Douglas et al., 1971a,b), this activity giving rise to 'coated vesicles' which promptly shed their coats to become smooth. The function of this process we believe is to effect a retrieval of 'granule' membrane and to conserve the cell surface. The phenomenon can be recognized in other cells and the term 'vesiculation' has been proposed (Douglas & Nagasawa, 1971, 1972).

Our evidence is as follows: (1) micropits or caveolae can be detected within exocytotic pits in the nerve ending (the depressions containing the extruded neurosecretory material) and these commonly show the 'coating' stigmatic of membrane vesiculation. Moreover, coated vesicles can be detected in the cytoplasm adjacent to such pits (Nagasawa et al., 1970; Douglas et al., 1971a). (2) In preparations fixed during acute stimulation coated vesicles are especially numerous

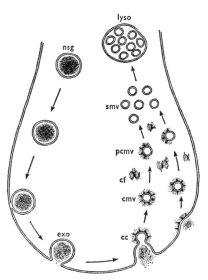

Fig. 2. *A diagram of the proposed mechanism of discharge of posterior pituitary hormones by exocytosis and the formation of smooth microvesicles ('synaptic vesicles') by vesiculation of exocytotic pits*

The neurosecretory granules (nsg) arrive from the perikaryon and discharge by exocytosis (exo). Coated micropits or caveolae (cc) then form in the exocytotic pit, round up and pinch off to form coated microvesicles (cmv) which shed coat fragments (cf) to yield first partially coated microvesicles (pcmv) then smooth ('synaptic') microvesicles (smv) which are probably incorporated into lysosomes. For references see the text.

and occur mingled with vesicles retaining only fragments of coating, characteristic alveolate figures of shed coats, and smooth 'synaptic' microvesicles (Douglas *et al.*, 1971*b*). (3) When neurohypophyses are exposed to horseradish peroxidase, a classical electron-microscopic marker of pinocytotic activity (it is a protein incapable of crossing intact membranes but readily seen in electron microscopy) this marker is subsequently found in both coated and smooth ('synaptic') microvesicles (Nagasawa *et al.*, 1971). The sequence we have proposed is depicted in Fig. 2. Several of the components of this sequence are discernible in invertebrate neurosecretory fibres (Normann, 1970; Bunt, 1969) and the mechanism may well be a general one. The scheme harmonizes not only with the evidence that synaptic vesicles increase in number after stimulation but with each of several other characteristics of the 'synaptic vesicle' population (Douglas *et al.*, 1971*a*; Douglas, 1974*b*). Clearly it provides an explanation for the long-enigmatic 'synaptic' vesicles and removes any need to entertain notions (see Bargmann, 1968) that they contain some transmitter substance participating in hormone release. From the present standpoint, however, the importance of the sequence lies in the conclusion that smooth microvesicles result from exocytosis and are thus indicative of prior exocytotic activity. This interpretation I shall buttress shortly by reference to a similar exocytosis–vesiculation sequence in chromaffin cells which once again offers certain technical advantages. Meanwhile I propose pursuing some further inferences we have drawn from this 'exocytosis–vesiculation sequence' which provide a convenient starting point for developing discussion of the question of exocytosis in ordinary nerves.

Inasmuch as vesiculation of exocytotic pits results in retrieval of membrane that originally surrounded the neurosecretory granule (see Fig. 2) one must consider the possibility that this membrane is reused for packaging secretory product. In the neurohypophysis we have considered this most unlikely since all packaging of hormone is performed in the remote perikarya where neurosecretory granules seem to be produced by the Golgi apparatus in the manner conventional for secretory cells (Sachs, 1969). However, we have suggested that such a similar tightly coupled exocytosis–vesiculation sequence (Fig. 3) resulting in retrieval of membrane with properties appropriate for storage and release of secretory product could occur in ordinary nerve endings and allow 'recycling' of synaptic vesicle membrane (Douglas *et al.*, 1971*a*; Douglas & Nagasawa, 1972; Nagasawa & Douglas, 1972). This possibility seemed to us attractive since these endings, unlike those of the neurosecretory fibres, synthesize large amounts of transmitter locally and receive a supply of preformed material by axonal transport from the perikaryon that falls far short of that needed to sustain ongoing activity. Moreover, there is strong evidence that locally synthesized material can be released only when incorporated into vesicles (Smith, 1972).

Now this scheme of 'recycling of synaptic vesicles' (Fig. 3) is a hypothetical one we arrived at by extrapolation from the neurohypophysial evidence (Douglas *et al.*, 1971*a*). What renders it particularly relevant to the question of exocytosis in nerves is that a comparable 'recycling' hypothesis has been arrived at independently and more directly by experiments on motor nerve endings by Heuser & Reese (1972, 1973). There has long been evidence of 'micropinocytosis' in nerve endings as indicated by the presence of coated pits and microvesicles and uptake

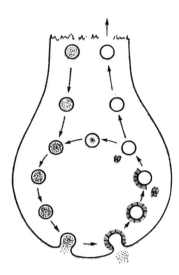

Fig. 3. *A diagram of the conjecture that a tightly coupled exocytosis–vesiculation sequence could operate in ordinary nerves to permit 'recycling' of synaptic vesicles*

The ending is supplied by synaptic vesicles at a slow rate from the perikaryon (downward arrow). These undergo exocytosis. The exocytotic pit so formed is, by analogy with events in the neurosecretory fibre (Fig. 2), coated and recaptured by vesiculation, loses its coat and, because it is the same membrane as that originally suited to transmitter incorporation and storage, acquires a fresh complement of transmitter and is 'recycled' until exhausted whereupon (upward arrow) it may suffer a lysosomal fate (after Douglas *et al.*, 1971*a*; see also Nagasawa *et al.*, 1971; cf. Ceccarelli *et al.*, 1973; Heuser & Reese, 1973).

of horseradish peroxidase. Moreover, horseradish peroxidase uptake has been noted to increase on stimulation. Heuser & Reese (1972, 1973), whose papers should be consulted for the earlier literature, have found, as have Pysh & Wiley (1972), that on vigorous stimulation there is a fall in the number of synaptic vesicles in cholinergic terminals and an expansion of the surface area of the terminal approximately corresponding to the area of the membrane of the 'lost' synaptic vesicles. Restoration of membrane area and the synaptic vesicle population then occurs, seemingly by membrane vesiculation, and horseradish peroxidase can be shown first in coated vesicles, and somewhat later in smooth 'synaptic' vesicles and disappears when stimulation is repeated. Although exocytosis has not been seen, it seems the only reasonable explanation for this sequence. Somewhat similar results have also been obtained by Ceccarelli *et al.* (1973). Although the evidence is indirect it is nevertheless an important new addition to the existing and controversial morphological evidence for exocytosis in ordinary nerve endings (see also Hubbard, 1970).

To complement this new morphological evidence there is gradually accumulating, for the adrenergic neurone, the 'kind of chemical evidence that has clarified events in the medullary chromaffin cell' (see reviews by Douglas, 1968*a*; Smith, 1971*a*, 1972; De Potter *et al.*, 1972). Through sensitive enzymic and immunological techniques release of noradrenaline is now known to occur along with other constituents of noradrenergic vesicles, dopamine β-hydroxylase and chromogranins which are both proteins presumably incapable of traversing intact membranes. Although there are formidable difficulties in establishing that these various substances are released in the proportions found in the vesicles (see review by De Potter *et al.*, 1972) the essential point is that considerable amounts do escape, and this is most simply explained by exocytosis.

It should perhaps be emphasized that besides the findings cited above, pharmacological analyses indicate that nerves, adrenergic or cholinergic, cannot release transmitter from the cytoplasm but only transmitter present within the vesicle fractions (Smith, 1972); and, of course, the familiar evidence that acetylcholine is discharged in multimolecular packages (quanta) also fits the exocytosis concept (Del Castillo & Katz, 1956; Katz, 1969), although as pointed out by Katz (1965) and Hubbard (1970) quantal discharge could be explained in other ways and these alternatives have been raised again with the recent discovery of subpopulations of quanta.

Although it is obvious that the evidence for the operation of exocytosis in ordinary nerves still falls far short of constituting proof it is certainly already substantial. That nerves should utilize exocytosis would be consistent with their common developmental origin with neurosecretory fibres and chromaffin cells and would account for the many parallels in their behaviour to ions and drugs (Douglas, 1968a).

Calcium influx necessary and sufficient

A requirement for calcium in the release of acetylcholine has been known for many years (Harvey & McIntosh, 1940) and when Hodgkin & Keynes (1957) found that impulses increased uptake of ^{45}Ca into squid giant axon they speculated whether 'a penetration of calcium at the nerve ending might not be one of the factors involved' in acetylcholine release (see also Birks & McIntosh, 1957). And the concept of a 'calcium receptor for acetylcholine release' somewhere within the terminal was later discussed in relation to the effects of barium which had then been shown capable of substituting for calcium (Douglas *et al.*, 1961). Such conjectures were, of course, based on the slenderest evidence capable of interpretation along different lines (see, e.g., Katz, 1962). They became much more attractive when calcium's role in stimulus–secretion coupling in the chromaffin cell began to be evident, for here—largely as a reflection of the simplicity and favourable nature of the preparation—the evidence for secretion mediated by calcium influx was much more extensive. Although the physiologial sequence in the chromaffin cell appeared to be: acetylcholine → membrane receptor activation → increased membrane permeability → influx of calcium → extrusion of catecholamines, it was evident that elevation of $[K^+]_0$ could substitute effectively for acetylcholine allowing the following sequence: excess of potassium → membrane depolarization → increased membrane permeability → influx of calcium → extrusion of catecholamines. The initial events here were recognized at once as corresponding to those in nerves and taken to indicate that the chromaffin cells, like neurons, their close developmental relatives, possess a voltage-dependent channel which opens with depolarization allowing entry of calcium down its electrochemical gradient (Douglas & Rubin, 1961, 1963; see also Douglas & Poisner, 1962; Douglas *et al.*, 1967a,b). When the isolated neurohypophysis was found to release posterior pituitary hormones on exposure to excess of potassium, and that this effect also required calcium, it was natural to extend the calcium-influx hypothesis to the terminals of the hypothalamo-neurohypophysial tract, and the specific suggestion was made that release of hormone is effected by 'arrival of impulses... which, by depolarizing the endings promote calcium influx ... and that the appearance of free calcium ions someplace in the endings then causes the release of stored hormone' (Douglas, 1963). This interpretation was soon reinforced by experiment showing that, as in the chromaffin cells, calcium was the only ion required (Douglas & Poisner, 1964a), accumulated in the preparation during stimulation (Douglas & Poisner, 1964b), and was essential for electrical stimulation (Mikiten & Douglas, 1965). About this time evidence began to build up, largely as a result of the elegant electrophysiological studies of Katz & Miledi (1965), which was to provide a remarkably clear picture of the involvement of calcium in transmitter release. By means of an ingenious focal calcium-electrode technique, Katz & Miledi (1965) had shown that the familiar effect of calcium deprivation in blocking impulse-evoked acetylcholine release was not attributable to any conduction failure in the terminals but involved the release process directly. Although they emphasized that 'the exact place of calcium action in the process of transmitter release remains unknown', they cited two possibilities: either that calcium acts by entering the terminals or that 'the change in the axon surface which leads to increased probability of quantal release requires the participation of calcium'. About 2 years later, after a further series of experiments of consummate design and technical brilliance involving the application of local depolarizing pulses to cholinergic terminals in concert with precisely timed pulses of calcium ions delivered by microelectrophoresis, they discarded the second scheme and envisioned 'the evoked release of quantal packets of acetylcholine proceeds in several stages the first of which is an inward movement through the axon membrane either of Ca^{2+} itself or a positively-charged calcium compound, CaR^+'. A splendid personal account of the evidence supporting the calcium-influx hypothesis of release of acetylcholine and other transmitters was presented by Katz in his Sherrington Lectures (Katz, 1969). Some other and more recent evidence will be found in reviews by Hubbard (1970) and Baker (1972). Among the many pieces of evidence are: (1) transmitter release is unaffected by tetrodotoxin which blocks voltage-dependent sodium-influx, but is inhibited by Mg^{2+} and Co^{2+}, ions blocking the voltage-dependent 'slow' calcium channel in invertebrate giant axonal membrane; (2) that ^{45}Ca accumulates in synaptosomes upon stimulation (Blaustein, 1971); (3) that transmitter secretion can be evoked by electrical stimulation of nerve terminals in an environment consisting essentially of $CaCl_2$; (4) that in the last-mentioned condition regenerative potentials can be obtained indicating that calcium is traversing the cell membrane and carrying inward current; and (5) that the efficacy of depolarization in inducing transmitter release is lost when membrane potential is driven far beyond zero and the inside rendered about 130mV positive with respect to the outside. Since this value corresponds roughly to the calculated calcium equilibrium potential, the observation fits nicely with the idea that passive downhill influx of calcium is the critical event.

That cholinergic nerves can be made to secrete in an environment consisting essentially of $CaCl_2$ (Katz & Miledi, 1969) is reminiscent of the behaviour of chromaffin cells (Douglas & Rubin, 1963). Similarly it has proved possible to elicit secretion of posterior pituitary hormones from neurohypophyses suspended in 'calcium–sucrose' media (Douglas & Sorimachi, 1971) and this effect, as in nerves (see Hubbard, 1970; Baker, 1972) and chromaffin cells (Douglas & Rubin, 1961, 1964a,b), is inhibited by ions such as Mg^{2+} and Co^{2+} known to block the 'slow' voltage-dependent calcium channel in invertebrate axon.

Unity of Events in Stimulus–Secretion Coupling in Neurons, Neurosecretory Fibres and Chromaffin Cells

The parallel behaviour of neurons, neurosecretory fibres and chromaffin cells is becoming increasingly well documented. It is thus fitting to re-emphasize the essential unity of events in the three systems to which attention was drawn some years ago (Douglas, 1966).

If one considers the chromaffin cell as the archetypical cell capable of secreting immediately in response to an imposed alteration at the cell surface whose recognizable components are: (1) permeability increase, (2) entry of Na^+ and Ca^{2+} and (3) depolarization, then the impulse-generating mechanism possessed by neurosecretory fibres and neurons can be regarded simply as stratagem evolved for telegraphing this same primitive pattern (the 'go' signal) to the remote secretory pole as illustrated in Fig. 4.

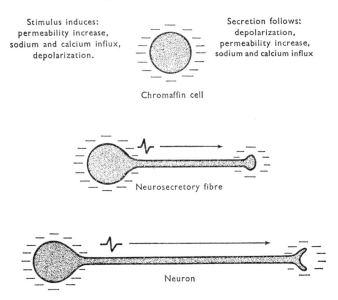

Fig. 4. *Unity of events in stimulus–secretion coupling in the three developmentally related cells; chromaffin cells, neurosecretory fibres of the hypothalamo-neurohypophysial tract and ordinary neurons*

In neurosecretory fibres and neurons stimulation sets up in the cell bodies the same basic pattern that suffices to initiate secretion directly in the chromaffin cell. In these two specialized secretory cells, however, this basic pattern, by virtue of electrical excitability, is converted into the impulse. But the impulse, once propagated to the terminals, imprints on them essentially the same pattern of depolarization and ion influx occurring at the perikarya or chromaffin cell. The impulse is in effect a device for telegraphing the basic primitive signal to secrete to secretory poles far removed from the point of stimulation (after Douglas, 1966).

Variations on the Calcium-Activated Exocytosis Theme

In marshalling evidence for the generalized concept of stimulus–secretion coupling I argued some years ago (Douglas, 1968a) that events discerned in the chromaffin cell would probably be applicable to others and that the essence of stimulus–secretion coupling in cells of widely different form and function was probably calcium-activated exocytosis. This view has since been sustained by experiments on a wide variety of cells, in reflexion more of the parsimony of Nature than the prescience of the writer. The β-cells of the endocrine pancreas and cells secreting various anterior pituitary hormones, for example, conform nicely to the chromaffin model: elevation of $[K^+]_0$ substituting for the physiological secretagogue—elevation in blood glucose or the appropriate hypothalamic-releasing hormone—calcium being essential for both, and barium itself acting as a secretagogue. Moreover, all these cells, like the chromaffin cell, seem to secrete by exocytosis. The response to excess of K^+ suggests that all these cells, as originally proposed for the chromaffin cell (Douglas & Rubin, 1961), possess a voltage-dependent calcium channel opening on depolarization to allow calcium ions to run inward down their electrochemical gradient. What I wish to draw attention to here is that some other cells do not follow this pattern so closely although they conform to the broader scheme of calcium-activated exocytosis. This and other variations are conveniently illustrated by the mast cell which offers us the advantage of a secretory activity that can be observed directly by light microscopy in the living cell and recorded by microcinematography, as will be demonstrated later.

When the function of calcium in chromaffin cell secretion was first becoming evident, it was already known that histamine release from sensitized mast cells on exposure to antigen was calcium-dependent and increased with increasing extracellular calcium concentration. Although a somewhat different interpretation of the calcium-dependence had been offered (see Mongar & Schild, 1962) the parallelism with events in the chromaffin cell led Rubin and I to suggest that calcium entry was occurring and mediating histamine release (Douglas & Rubin, 1961). On this view (see also Douglas, 1968a, 1970) the interaction of antigen with membrane-bound antibody would resemble the interaction of acetylcholine and its chromaffin-cell receptor in that its functional importance would reside in an increased membrane permeability to calcium. Later the more specific suggestion was made that we were again dealing with calcium-activated exocytosis (Douglas, 1968a). The calcium-influx hypothesis gains support from intracellular recording with microelectrodes showing that during the hypersensitivity reaction mast cells suspended in a sodium-free medium depolarize and this depolarization increases with increasing $[Ca^{2+}]_0$ (Douglas & Kanno, 1969). With the now quite beautiful morphological evidence for exocytosis in mast cells (Bloom & Chakravarty, 1970; Röhlich et al., 1971; Lagunoff, 1973) this scheme is all the more attractive (see 'Note added in proof').

Secretion to excess of potassium and voltage-dependent calcium influx

Superficially, then, the mast cell would seem to conform to the chromaffin cell model. However, unlike the chromaffin cell, the mast cell does not secrete when the extracellular potassium concentration is raised. This can be inferred from observations made many years ago on lung tissue (Schild, 1936) and under direct visual observation elevating extracellular potassium to 60 or 150mM (with appropriate decrease in Na^+ to maintain osmotic pressure) causes none of the characteristic signs of mast cell secretion visible in light microscopy (Cochrane & Douglas, 1974) despite the fact the mast cells have a positive resting potential outside of about 10mV which falls on increasing extracellular potassium concentration (Douglas & Kanno, 1969; Kanno et al., 1973; W. W. Douglas & T. Kanno, unpublished work). One interpretation would be that the mast cell membrane possesses calcium channels that may be opened by the hypersensitivity reaction but which are unaffected by potassium i.e., are not voltage-dependent. Comparable behaviour, with a possibly similar explanation, can be recognized in other cells. For example, protein secretion from salivary glands, and the exocrine pancreas, which is calcium-dependent (Douglas & Poisner, 1962; Hokin, 1966), cannot be elicited by increasing $[K^+]_o$, provided secondary effects mediated through the nerves are avoided (Schramm, 1968; Argent et al., 1971) although transmembrane potential falls sharply (Dean & Matthews, 1972; Kanno, 1972). Likewise, but in contrast with all other adenohypophyseal cells, prolactin-secreting cells are not stimulated by elevated $[K^+]_o$, although secretion is again calcium-dependent (see Geschwind, 1971).

Different sources of calcium: mobilization and influx

The second interesting feature of the mast cell is its seeming capacity to utilize cellular calcium stores to sustain exocytosis. Whereas mast cell secretion evoked by antigen is readily abolished by lowering $[Ca^{2+}]_o$, as is the chromaffin cell's response to acetylcholine, secretion evoked by the polyamine 48-80, which likewise involves exocytosis (Horsfield, 1965; Röhlich et al., 1971), has been elicited in calcium-free environments even in the presence of chelating agents (Diament & Krüger, 1967; Uvnäs & Thon, 1961). On first view this result might seem to call into question the whole concept of calcium as a critical coupling agent in exocytosis. My colleagues and I have confirmed that mast cells release histamine (Douglas & Ueda, 1973) and extrude granules (Cochrane & Douglas, 1974) in characteristic exocytotic fashion in response to polyamine 48-80 after brief exposure to calcium-free media but find they no longer respond when incubated with 2mM-EDTA for 3h. Such EDTA-treated cells rapidly regain their responsiveness to polyamine 48-80 when calcium is re-introduced into the medium (Fig. 5). By contrast magnesium is ineffective. Clearly then the response to polyamine 48-80 is not independent of calcium as might have been supposed from earlier work but apparently draws on some cellular source of calcium not rapidly depleted by calcium-free media or brief exposure to EDTA. The point here is that there are other ways in which calcium can participate in stimulus–secretion coupling besides passive influx down an electrochemical gradient. A decade ago it was suggested that 'mobilization' of membrane-bound calcium might contribute to the secretory response (Douglas, 1963). The responses of the mast cell to polyamine 48-80 could be interpreted in a similar vein but the critical source of calcium could equally well lie elsewhere.

Once again comparable behaviour can be found in other cells. When the calcium-dependence of secretion of protein from salivary glands was first observed it was

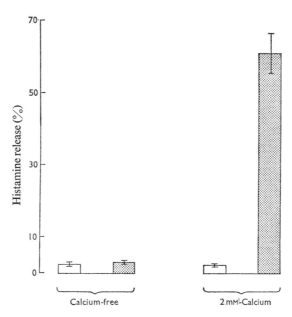

Fig. 5. *Calcium-dependent nature of histamine release (exocytosis) induced by polyamine* 48-80 *in the mast cell*

Mast cells isolated from the peritoneal cavity of Sprague–Dawley rats (about 200g) were incubated at 37°C for 3h in calcium-free Locke solution containing 2mM-EDTA. Thereafter they were washed and resuspended in calcium-free Locke solution (left hand records) or Locke solution with 2mM-calcium (right hand records) for 25min. During the last 10min portions were exposed to polyamine 48-80 (1 μg/ml), ▨, as indicated. In the presence of calcium, polyamine 48-80 released about 60% of total histamine. In the absence of calcium it had no histamine-releasing activity. The unhatched columns show the degree of spontaneous histamine release in the control samples without polyamine 40-80. The S.E.M. are indicated; $n = 3$ (Douglas & Ueda, 1973).

noted that secretion did not fail as promptly on removal of extracellular calcium as did secretion from the adrenal medulla and that it was much less inhibited by increasing extracellular magnesium known at the time to decrease calcium influx in axons. Although it was felt that these effects did not necessarily exclude the possibility that calcium entry is the factor eliciting the secretory response they encouraged speculation at the time of alternative explanations of the role of calcium (Douglas & Poisner, 1962). At this present meeting Case *et al.* (1973) provide evidence that pancreatic secretagogues, cholecystokinin-pancreozymin and acetylcholine, may initiate secretion by releasing calcium from intracellular stores.

The cellular sources of calcium participating in mast cell and exocrine secretions have yet to be identified and clearly there may be other secretory cells that utilize similar or related mechanisms. It is possible that the sources of calcium utilized in stimulus–secretion coupling are as diverse as those participating in excitation–contraction coupling where the pattern of utilization of extracellular and cellular calcium varies widely between smooth, cardiac and skeletal muscles.

Serial fusion of granules: 'compound' versus 'simple' exocytosis

The third point illustrated by the mast cell is that the process of secretion commonly referred to as exocytosis exists in different forms. This is not generally appreciated but is, I believe, of critical importance for arriving at an understanding of the basic biochemical events involved in the fusion–fission reaction central to the exocytotic event. Everyone recognizes that exocytosis is a fusion–fission response involving interaction of two membranes: that of the cell surface and that surrounding the secretory granule. What is not generally recognized, however, is that the membrane composing the cell surface is not always plasma membrane in the conventional sense (i.e., membrane not merely delimiting the cell surface but having unique characteristics) but sometimes the membrane of a secretory granule or granules incorporated into the cell surface as a result of previous exocytotic events. In the mast cell stimulated with polyamine 48-80 this process of 'serial fusion' of granules has been demonstrated by electron-microscopic methods (Horsfield, 1965; Röhlich *et al.*, 1971) and may be readily seen with Nomarski optics, as was seen in the film shown. The serial fusions result in formation of large intracellular cisternae which open to the cell surface through more or less large pores, the site of the initial exocytotic event. When such serial fusion–fission is occurring it is quite evident that granule membranes are interacting with each other. This behaviour contrasts with that of the chromaffin cell where exocytosis seems to involve only one chromaffin granule at a time and where the fusion–fission reaction can reasonably be supposed to involve an interaction of plasmalemma proper and granule membrane (Benedeczky & Smith, 1972; Diner, 1967;

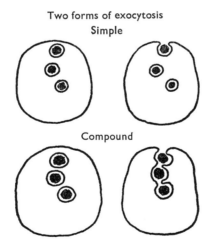

Fig. 6. *The two forms of exocytosis: simple and compound*

In simple exocytosis a single granule approaches the plasma membrane, melds with it and, after rupture of the conjoined plasma and granule membranes, voids its contents into the extracellular space. The essential fusion–fission reaction involves two functionally and chemically distinct membranes: plasmalemma and granule membrane. In compound exocytosis the initial event is as above but thereafter other granules meld with the first and undergo the fusion–fission reaction. In these subsequent events the membranes involved are thus similar: the interaction is between granule membranes. The functional implications of the two forms of exocytosis are discussed in the text.

Nagasawa & Douglas, 1972). Once again behaviour paralleling that of the mast cell is found in other cell types. For example serial fusion of granules is abundant in salivary glands (Amsterdam *et al.*, 1972) and in adenohypophysial cells (Farquhar, 1971; De Virgilis *et al.*, 1968). Because of the important functional implications of the two different categories of exocytosis I propose we recognize this by providing them with different names: exocytosis involving plasmalemma and a single granule might be referred to as 'simple exocytosis' and that involving serial melding and rupture of granules as 'compound exocytosis' (Fig. 6).

The variations discussed in this section—differences in responsiveness to elevation of $[K^+]_0$ or sources of calcium utilized and employment of 'simple' or 'compound' exocytosis—may correspond to the different embryological origins of the cells. The possession, by the chromaffin cells, of a voltage-dependent calcium-influx mechanism, in addition to that directly operated by acetylcholine, and the dependence of that cell on $[Ca^{2+}]_0$ has from the beginning been associated with its common developmental origins with neurons; and it is noteworthy that the cells whose secretion is most intimately regulated by $[Ca^{2+}]_0$, the calcitonin-secreting C cells of the thyroid, have a neural crest origin (Pearse & Polak, 1971a; Le Douarin & Le Lievre, 1970); and indeed that a large number of endocrine polypeptide cells may also share this origin (Pearse & Polak, 1971b). A comparative study is clearly in order but is complicated by the uncertain origin of many endocrine cells.

Aspects of Exocytosis Illustrated by Microcinematography of Mast Cells

The secretory response of mast cells to polyamine 48-80 (or antigen in the case of sensitized mast cells) is one of the most remarkably dynamic cellular responses that it is possible to see and several authors have made films of the process (e.g., Horsfield, 1965).

With the evidence now strong that polyamine 48-80 induced secretion conforms to the general pattern of calcium-dependent exocytosis (see the preceding section) this approach has acquired added significance and Dr. J. Nagasawa and I have applied high-speed microcinematography (up to 500 frames/s) in an attempt to better resolve rapid individual exocytotic events and arrive at a clearer understanding of the dynamics of exocytosis. The resulting film, along with footage shot at slower speeds, will illustrate some aspects of exocytosis in the living cell. Segments of the film are shown in Plate 1.

At this point in the presentation a short film was shown of living isolated peritoneal mast cells of the rat, some showing 'spontaneous' exocytosis others responding to polyamine 48-80. The following points were made. (1) In the normal mast cell secretory granules appear quite stationary. To some this may seem surprising since there is a rather widely held view that secretory granules or vesicles are normally in Brownian motion, and this assumption has been incorporated into theoretical discussions of secretory events in e.g. neurons (Katz, 1969) and chromaffin cells (Banks, 1970; Matthews, 1970). But clearly secretory granules in the living mast cell show no such activity. They appear immobile and retain their station and relation one to the other over long periods. The lively dance of granules

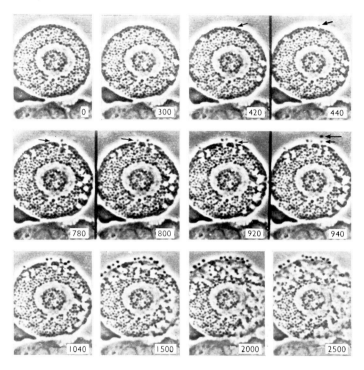

EXPLANATION OF PLATE I

Segments of a microcinematographic film showing the secretory response (compound exocytosis) of an isolated peritoneal rat mast cell to polyamine 48-80

The cell is viewed under phase contrast in Locke solution at 37°C. The histamine-containing granules appear as small dark balls. The appearance of the cell before polyamine 48-80 treatment is indicated in the top left hand frame (0). The drug was then added to a final concentration of about 100 μg/ml and the response followed while filming at 50 frames/s. Representative frames are shown at various times thereafter. The numbers give the time in ms after the first visually detectable change. Note: (1) the serial ejection of two pairs of granules in the records indicated by the arrows, (2) the formation of lacunae resulting from multiple, granule–granule, 'compound' exocytosis (J. Nagasawa & W. W. Douglas, unpublished work).

in Brownian motion can, however, be readily observed when the cell is maintained for long periods under adverse metabolic conditions and when the response to polyamine 48-80 is lost. Zymogen granules in normal pancreatic cells are equally immobile (T. Kanno, personal communication). Moreover, my own observations on living chromaffin and adenohypophysial cells obtained by tissue disaggregation (see Douglas et al., 1967a) lead to the same conclusion. In these endocrine cells granules are small and at the limits of resolution by light microscopy but the stability of the pattern of the granules within the cell is quite evident and contrasts sharply with the Brownian motion which is again readily induced in these cells by various procedures causing cell death.

(2) Exposure of mast cells to polyamine 48-80 causes, within seconds, a rapid expulsion of large numbers of granules. Precise timing of the essential exocytotic event—the fusion–fission reaction causing ejection to the exterior of granule content—is impossible, for light microscopy cannot resolve membranes, but from the motion of individual granules evident at 500 frames/s and from frame by frame analysis it seems granule expulsion is completed within about 6–8 ms, and obviously this only provides an upper value. Given the speed of the reaction it is not surprising that conventional fixation methods of electron microscopy have so far failed to provide details of the critical fusion–fission process. Or indeed, that images of exocytosis are rarely captured in many cells such as the chromaffin cells of the cat adrenal gland or mammalian neurosecretory systems despite the persuasive chemical evidence for exocytosis.

(3) From some sequences it seems that mast cells shrink a second or so before granule expulsion begins. This could be simply an artifact resulting from rounding up of flattened cells, but considered along with the sometimes vigorously expulsive discharge of granules it lends some support to conjecture contractile elements are involved in stimulus–secretion coupling (see below).

(4) The serial fusion of granules ('compound exocytosis') can be followed as a progressive involvement of granules beginning at the cell surface and rapidly penetrating deeper. The swelling and loss of optical density of discharged granules is characteristic, as is the formation of large cisternae containing them. Cisternal openings to the cell surface are sometimes clearly signalled by the outward passage of several granules in Indian file. Such focal extrusion of a series of granules is, at least to this observer, reminiscent of defaecation in the goat. Others might find less earthy analogies (Plate 1).

Possible Involvement of Calcium in 'Vesiculation Coupled to Exocytosis'

As discussed above, the calcium-mediated process set in motion in neurohypophysial terminals, chromaffin cells (and probably also in neurons and various other cells) involves not merely extrusion of secretory product but recapture of granule membrane from the cell surface by vesiculation. Quantitative evidence of the truly preferential nature of this membrane vesiculation has been obtained from the chromaffin cell where the incidence of coated caveolae within exocytotic pits was found to be about five times higher than in adjacent cell membrane (Douglas & Nagasawa, 1972). The functional aspects of this exocytosis–vesiculation sequence have been considered already from the standpoint of vesicle 'recycling'.

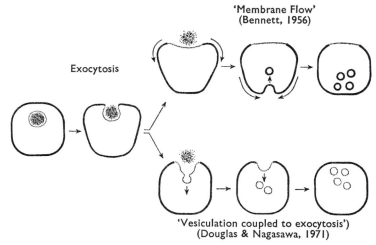

Fig. 7. *A comparison between two suggested mechanisms for countering the cell surface-expanding effect of exocytosis*

Above is the classical scheme after Bennett (1956) involving a conveyor belt-like 'membrane flow' where exocytosis is countered by pinocytotic activity. Note the cell surface area is maintained but is now 'diluted' with secretory granule membrane (dotted line). Below is the alternative involving a micropinocytosis-like activity superimposed on the exocytotic pit itself and referred to as 'vesiculation coupled to exocytosis' (Douglas & Nagasawa, 1971). Note this restores not only the area of the cell surface but its chemical composition. It also retrieves 'granule' membrane.

The mechanism also provides for conservation of cell surface area and chemical composition (Douglas & Nagasawa, 1971) as depicted in Fig. 7 where it is compared with the classical membrane-flow mechanism.

What concerns us here is whether calcium is intimately involved in both components of the sequence, vesiculation as well as exocytosis. Dr. Nagasawa and I have been attracted to this possibility by several considerations. It is evident that vesiculation involves conversion of a surface of large curvature into spheres of smaller radius and thus must require considerable force. This is no doubt provided by the mysterious 'coating' (Roth & Porter, 1964; Kanaseki & Kadota, 1969). The problem of chemo-mechanical transduction leads naturally to analogies with contractile systems and to the suspicion that calcium may participate. And this suspicion becomes stronger as experiments diminish the attractiveness of alternatives.

The essence of the matter is that for coating and vesiculation to occur some change must take place in the membrane of the secretory granule when it is incorporated into the cell surface. Dr. Nagasawa and I have been exploring two possibilities: that the change is imposed by the different electrical field or the different ionic environment (Douglas & Nagasawa, 1972). There is no reason to believe that any significant potential gradient exists across the secretory granule membrane as it lies in the cytoplasm. But once incorporated in the cell surface, granule membrane will necessarily be subjected to the sizeable potential gradient across the cell membrane. In the chromaffin cell this amounts to some 30mV. When imposed across a membrane 10 nm (100 Å) thick this yields a value of 30 000 V/cm, a very respectable field that conceivably could reorient charged molecules in the membrane to provide the stimulus for coating. To test this we have incubated hamster adrenal glands in a calcium-free environment, exposed them to elevated potassium concentrations to depolarize the cells, and then evoked secretion by introducing calcium. Electron micrographs showed abundant exocytosis with evidence of superimposed vesiculation. Unless there is some technical fault to this experiment some event other than electrical potential change must initiate vesiculation. This heightens interest in the ions and the possibility that vesiculation results from exposure of granule membrane to the relatively high extracellular concentrations of sodium and calcium. Sodium does not seem to be involved, since vesiculation can be seen in glands incubated in sodium-free media. Attention thus focuses on calcium which, it will be recalled, is present extracellularly in a concentration several orders higher than that found intracellularly (Baker, 1972). Unfortunately a test of calcium's involvement in vesiculation has so far eluded us since the ion is essential for exocytosis. Moreover, one cannot ignore the alternative or additional possibility that vesiculation is induced or favoured by granule contents such as ATP, chromogranins, or neurophysin extruded along with the characteristic hormonal products (Douglas & Nagasawa, 1972). All that can be said at present is that calcium remains one of the more attractive putative 'membrane vesiculating' agents.

Calcium and 'Contractile' Hypothesis of Secretion

The parallels between excitation–contraction coupling and stimulus–secretion coupling which became evident in the early experiments on the adrenal (Douglas & Rubin, 1961) extend far beyond calcium requirement, capacity of barium and strontium to substitute, ability of magnesium to inhibit, and mimicry of the physiological stimulus by elevation of extracellular potassium. It is the rule rather than the exception that any manipulation of the ionic environment or exposure to drugs stimulating or inhibiting secretion can be shown to have a corresponding effect on contraction in one or another type of muscle, smooth, cardiac or skeletal; and, moreover, that stimulatory and inhibitory influences can often be related to elevated and depressed inward calcium fluxes or intracellular free calcium ion concentration. This parallelism led naturally to the comment that 'perhaps Nature in her wisdom has found calcium entry or mobilization a convenient way of initiating secretion or contraction . . . and that it may further the efforts of those engaged in one or another of these fields to see our work within the framework of the more general concept of stimulus–response coupling' (Douglas, 1966). Attempts to find a 'contractile' basis for secretion, or involvement of the same sort of basic reactions as are involved in contraction, have so far provided no definitive evidence but have spawned a number of highly speculative schemes. Broadly speaking there have been two lines of inquiry. The first revolving around granule ATPase activity. The second being of a more frankly contractile nature and concerned with microfilaments and microtubules.

Granule ATPase as a factor

It has been suggested that activation of ATPase activity in the secretory granule membrane leads in some manner to the molecular rearrangements permitting the melding and rupture of granule and cell membranes central to exocytosis (Poisner & Trifaró, 1967; Poisner & Douglas, 1968; Trifaró & Poisner, 1967). Mg^{2+}-activated ATPase activity is present in secretory granule membranes from a variety of cells (Douglas, 1968a,b; Poisner, 1970) and recent evidence of enhancement of activity by low concentrations of calcium (Izumi et al., 1971) has encouraged further speculation along these lines (see Fig. 4 in Douglas, 1973).

In a further development of the theme that draws the parallel with contraction ever closer, Berl et al. (1973) suggested that $Mg^{2+}-Ca^{2+}$-activated ATPase of synaptic vesicles reflects a myosin-like activity, whereas plasmalemma has actin-like properties. In their view calcium 'triggers' interaction between the two components and causes the fusion–fission reaction of exocytosis. How far this line of experiment will develop remains to be seen, but a word of caution seems in order. Although it is undoubtedly attractive to see exocytosis as resulting from some complementary properties of granule membrane and plasma membrane this approach may be misleading, certainly there is a difficulty in applying this concept to exocytosis in general for it fails to account for 'compound' exocytosis where, except for the initial event, the reactions are between adjacent granules (Fig. 6). Perhaps incorporation confers on the granule membrane the capacity to behave as plasmalemma and permit fusion–fission with the nearest underlying granule. Conceivably the factors conferring this property could be the same as those discussed in regard to vesiculation (see the previous section), namely exposure of the granule membrane to the electrical potential across the cell surface or to the extracellular ionic environment (high Ca^{2+}?). I find it interesting to compare this inward propagation of the fusion–fission reaction, which clearly involves a propagated membrane disturbance, with the propagated wave of altered membrane characteristics that underlies impulse transmission in electrically excitable tissue. Here is yet another exciting area for study.

Microtubules and microfilaments: colchicine and cytochalasin

A second set of 'contractile' hypotheses involves subcellular elements with putative contractile functions, the microtubules and microfilaments. Some years ago Lacy et al. (1968), proposed that in some illdefined manner such elements might propel the secretory granule into exocytotic activity and obviously such a frankly contractile scheme offers another possible basis for involvement of calcium.

The bulk of evidence comes from the effects of two types of pharmacological agents: on the one hand colchicine and related substances capable of disrupting microtubules, and on the other the mould metabolite cytochalasin supposed to interact in analogous fashion with microfilaments (Wessells et al., 1971). Both types of drugs interfere with various secretory systems. However, as is unfortunately true of many pharmacological analyses, interpretation is complicated by the unspecificity of the agents. For example, one of the most striking examples of inhibition of secretion produced by colchicine, inhibition of medullary secretion in response to cholinergic stimuli (Poisner & Bernstein, 1969), seems not to result from any direct effect on exocytosis, for responses to other secretagogues are little impaired, but to reflect an action on some early link in stimulus–secretion coupling

(Douglas & Sorimachi, 1972b; Trifaró et al., 1972). Moreover, both colchicine and vinblastine, in the high concentrations used in the secretory studies, inhibit functions quite unrelated to exocytosis, e.g. uptake of transmitters (Nicklas et al., 1973) and initiation of nerve impulses (Trifaró et al., 1972).

Effects of cytochalasin on secretions of various sorts are also difficult to interpret. In at least some instances there are grounds for suspecting the effects are not due to direct actions on microfilaments but are indirectly mediated by some metabolic action interfering with the energy supply for secretion. Cytochalasin is capable of blocking uptake of various hexoses including glucose (Kletzien et al., 1972; Mizel & Wilson, 1972; Zigmond & Hirsch, 1972). It strongly depresses acetylcholine output from cholinergic nerves only when the nerves are fatigued; and inhibition is then countered, in part, by pyruvate (Nakazato & Douglas, 1973). And its effect on growth hormone secretion is also correlated with metabolic status (McPherson & Schofield, 1972). Still more striking evidence has come from the mast cell where the energy requirement for secretion can be met fully by oxidation of endogenous substrates or, when this is blocked, by glycolysis alone provided extracellular glucose is present. Mrs. Y. Ueda and I have found that polyamine 48-80-induced histamine secretion is influenced little or not at all by cytochalasin when mast cells are maintained in normal metabolic conditions, but is powerfully inhibited by cytochalasin, even in much lower concentrations, when the response is rendered dependent on extracellular glucose by exposure to dinitrophenol. The preferential inhibitory effect of cytochalasin on histamine secretion when glucose uptake is rate-limiting can be duplicated by the inhibitor 2-deoxyglucose or simply by removing glucose from the incubation medium. It probably results from block of glucose uptake (Douglas & Ueda, 1973, and unpublished work). It thus seems that exocytosis, at least in the mast cell, is not directly affected by cytochalasin and therefore cannot involve critically any cytochalasin-sensitive element, microfilaments or any other.

In the light of this evidence and the universal need for energy in secretion, one must treat with reserve any claims that microfilaments are involved in exocytosis that rest entirely on effects of cytochalasin.

Concluding Remarks

In conclusion, the concept that calcium has a central role in stimulus–secretion coupling and serves to link stimulus to exocytotic response can now be supported by still more extensive and varied information than was possible only a few years ago; and there has been a correspondingly heightened interest in attempts to define calcium's function in exocytosis. Several promising leads have been uncovered, but it is too soon to predict with confidence whether any of these will provide the answer.

Note Added in Proof

The concept of calcium-activated exocytosis has recently been further strengthened by the demonstration that mast cells extrude granules when calcium is injected intracellularly (Kanno et al., 1973) or introduced extracellularly after 'priming' with calcium ionophores (Cochrane & Douglas, 1974).

The work by the author and his colleagues referred to in this paper was supported by Grants from the United States Public Health Service.

References

Amsterdam, A., Ohad, I. & Schramm, M. (1972) *J. Cell Biol.* **41**, 753–773
Argent, B. E., Case, R. M. & Scratcherd, T. (1971) *J. Physiol. (London)* **216**, 611–624
Baker, P. (1972) *Progr. Biophys. Mol. Biol.* **24**, 177–223
Banks, P. (1967) *J. Physiol. (London)* **193**, 631–637
Banks, P. (1970) in *Calcium and Cellular Function* (Cuthbert, A. W., ed.), pp. 148–162, St. Martins Press, New York
Banks, P. & Helle, K. (1965) *Biochem. J.* **97**, 40 c
Banks, P., Biggins, R., Bishop, R., Christian, B. & Currie, N. (1969) *J. Physiol. (London)* **200**, 797–805
Bargmann, W. (1968) *Handb. Exp. Pharmacol.* **13**, 1–39
Benedeczky, I. & Smith, A. D. (1972) *Z. Zellforsch. Mikrosk. Anat.* **124**, 367–386
Bennett, H. S. (1956) *J. Biophys. Biochem. Cytol.* **2**, Suppl. 99–103
Berl, S., Puszkin, S. & Nicklas, W. J. (1973) *Science* **179**, 441–446
Birks, R. I. & MacIntosh, F. C. (1957) *Brit. Med. Bull.* **13**, 157–161
Blaustein, M. P. (1971) *Proc. Nat. Acad. Sci. U.S.* **69**, 2241–2245
Bloom, G. D. & Chakravarty, N. (1970) *Acta Physiol. Scand.* **78**, 410–419
Bunt, A. H. (1969) *J. Ultrastruct. Res.* **28**, 411–421
Burton, A. M., Forsling, M. & Martin, M. J. (1971) *J. Physiol. (London)* **217**, 23 p–24 p
Case, R. M., Clausen, T. & Smith, R. K. (1973) *Biochem. Soc. Trans.* **1**, 857–858
Ceccarelli, B., Hurlbut, W. P. & Mauro, A. (1973) *J. Cell Biol.* **57**, 499–524
Cheng, K. W. & Friesen, H. G. (1970) *Metabolism* **19**, 876–890
Cochrane, D. E. & Douglas, W. W. (1974) *Proc. Nat. Acad. Sci. U.S.*, in the press
Dean, P. M. & Matthews, E. K. (1972) *J. Physiol. (London)* **226**, 1–13
de Duve, C. (1963) in *Lysosomes: Ciba Found. Symp.* p. 126
Del Castillo, J. & Katz, B. (1956) *Progr. Biophys.* **6**, 121–170
De Potter, W. P., Chubb, I. W. & De Schaepdryver, A. (1972) *Arch. Int. Pharmacodyn. Ther.* Suppl. **196**, 258–287
De Robertis, E. D. P. & Sabatini, D. D. (1960) *Fed. Proc. Fed. Amer. Soc. Exp. Biol.* **19** (Part II, Suppl. 5), 70–73
De Robertis, E. D. P. & Vaz Ferreira, A. (1957) *Exp. Cell Res.* **12**, 568–574
De Robertis, E. D. P., Nowinski, W. W. & Saez, F. A. (1965) in *Cell Biology*, 4th edn., pp. 432–433, W. B. Saunders, Philadelphia and London
De Virgilis, G., Meldolesi, J. & Clementi, F. (1968) *Endocrinology* **83**, 1278–1284
Diamant, B. & Krüger, P. G. (1967) *Acta Physiol Scand.* **71**, 291–302
Diner, O. (1967) *C. R. Acad. Sci. Ser. D* **265**, 616–619
Douglas, W. W. (1963) *Nature (London)* **197**, 81–82
Douglas, W. W. (1966) *Proc. Int. Wenner-Gren Symp. Stockholm*, pp. 267–290, Pergamon, London
Douglas, W. W. (1967) *Int. Symp. Neurosecretion 4th* pp. 178–190
Douglas, W. W. (1968a) *Brit. J. Pharmacol.* **34**, 451–474
Douglas, W. W. (1968b) *Proc. Int. Cong. Physiol. Sciences 24th*, vol. 6, pp. 63–64
Douglas, W. W. (1970) in *The Pharmacological Basis of Therapeutics* (Goodman, L. S. & Gilman, A., eds.), 4th edn., pp. 628–629, Macmillan, New York
Douglas, W. W. (1973) *Progr. Brain Res.* **39**, 35–39
Douglas, W. W. (1974a) in *Handb. Physiol. Sect.* 7 (Blaschko, H. & Smith, A. D., eds.), in the press
Douglas, W. W. (1974b) in *Handb. Physiol. Sect.* 7 (Sawyer, W. H., ed.), in the press
Douglas, W. W. & Kanno, T. (1967) *Brit. J. Pharmacol.* **30**, 612–619
Douglas, W. W. & Kanno, T. (1969) *Abstr. Proc. Int. Congr. Pharmacol. 4th*, p. 347
Douglas, W. W. & Nagasawa, J. (1971) *J. Physiol. (London)* **218**, 94 p–95 p
Douglas, W. W. & Nagasawa, J. (1972) *Proc. Can. Fed. Biol. Soc.* **15**, 757
Douglas, W. W. & Poisner, A. M. (1961) *Nature (London)* **192**, 1299
Douglas, W. W. & Poisner, A. M. (1962) *J. Physiol. (London)* **162**, 385–392
Douglas, W. W. & Poisner, A. M. (1963) *J. Physiol. (London)* **165**, 528–541
Douglas, W. W. & Poisner, A. M. (1964a) *J. Physiol. (London)* **172**, 1–18
Douglas, W. W. & Poisner, A. M. (1964b) *J. Physiol. (London)* **172**, 19–30
Douglas, W. W. & Poisner, A. M. (1966a) *J. Physiol. (London)* **183**, 236–248

Douglas, W. W. & Poisner, A. M. (1966b) *J. Physiol. (London)* **183**, 249–256
Douglas, W. W. & Rubin, R. P. (1961) *J. Physiol. (London)* **159**, 40–57
Douglas, W. W. & Rubin, R. P. (1963) *J. Physiol. (London)* **167**, 288–310
Douglas, W. W. & Rubin, R. P. (1964a) *J. Physiol. (London)* **175**, 231–241
Douglas, W. W. & Rubin, R. P. (1964b) *Nature (London)* **203**, 305–307
Douglas, W. W. & Sorimachi, M. (1971) *Brit. J. Pharmacol.* **42**, 647P
Douglas, W. W. & Sorimachi, M. (1972a) *Brit. J. Pharmacol.* **45**, 143P–144P
Douglas, W. W. & Sorimachi, M. (1972b) *Brit. J. Pharmacol.* **45**, 129–132
Douglas, W. W. & Ueda, Y. (1973) *J. Physiol. (London)* **234**, 97P–98P
Douglas, W. W., Lywood, D. W. & Straub, R. W. (1961) *J. Physiol. (London)* **154**, 39P–40P
Douglas, W. W., Poisner, A. M. & Rubin, R. P. (1965) *J. Physiol. (London)* **179**, 130–137
Douglas, W. W., Kanno, T. & Sampson, S. R. (1967a) *J. Physiol. (London)* **188**, 107–120
Douglas, W. W., Kanno, T. & Sampson, S. R. (1967b) *J. Physiol. (London)* **191**, 107–121
Douglas, W. W., Nagasawa, J. & Schulz, R. A. (1971a) *Mem. Soc. Endocrinol.* **19**, 353–378
Douglas, W. W., Nagasawa, J. & Schulz, R. A. (1971b) *Nature (London)* **232**, 340–341
Farquhar, M. G. (1961) *Trans. N.Y. Acad. Sci.* **23**, 346–351
Farquhar, M. G. (1971) *Mem. Soc. Endocrinol.* **19**, 79–124
Fawcett, C. P., Powell, A. E. & Sachs, H. (1968) *Endocrinology* **83**, 1299–1310
Geschwind, I. I. (1971) *Mem. Soc. Endocrinol.* **19**, 221–232
Harvey, A. M. & MacIntosh, F. C. (1940) *J. Physiol. (London)* **97**, 408–416
Heuser, J. E. & Reese, T. S. (1972) *Anat. Rec.* **172**, 329
Heuser, J. E. & Reese, T. S. (1973) *J. Cell Biol.* **57**, 315–344
Hodgkin, A. L. & Keynes, R. D. (1957) *J. Physiol. (London)* **253**–281
Hokin, L. E. (1966) *Biochim. Biophys. Acta* **115**, 219–221
Horsfield, G. (1965) *J. Pathol. Bacteriol.* **90**, 599–605
Hubbard, J. I. (1970) *Progr. Biophys. Mol. Biol.* **21**, 33–124
Izumi, F., Oka, M. & Kashimoto, T. (1971) *Jap. J. Pharmacol.* **21**, 739–746
Kanaseki, T. & Kadota, K. (1969) *J. Cell. Biol.* **42**, 202–220
Kanno, T. (1972) *J. Physiol. (London)* **226**, 353–371
Kanno, T., Cochrane, D. E. & Douglas, W. W. (1973) *Can. J. Physiol. Pharmacol.* in the press
Katz, B. (1962) *Proc. Roy. Soc. Ser. B* **155**, 455–477
Katz, B. (1965) *Proc. Int. Congr. Physiol. 23rd* pp. 110–121
Katz, B. (1969) in *The Sherrington Lectures*, pp. 1–60, C. C. Thomas, Springfield
Katz, B. & Miledi, R. (1965) *Proc. Roy. Soc. Ser. B* **161**, 496–503
Katz, B. & Miledi, R. (1969) *J. Physiol. (London)* **203**, 689–706
Kirshner, N. & Kirshner, G. (1971) *Phil. Trans. Proc. Roy. Soc. London Ser. B* **261**, 279–289
Kirshner, N. & Viveros, O. H. (1970) *Bayer Symp. 2nd* pp. 78–88
Kletzien, R. F., Perdue, J. F. & Springer, A. (1972) *J. Biol. Chem.* **247**, 2964–2966.
Krisch, B., Becker, K. & Bargmann, W. (1972) *Z. Zellforsch.* **123**, 47–54
Lacy, P. E., Howell, S. L., Young, D. A. & Fink, C. J. (1968) *Nature (London)* **219**, 1177–1179
Lagunoff, D. (1973) *J. Cell Biol.* **57**, 252–259
Le Douarin, N. & Le Lievre, C. (1970) *C. R. Acad. Sci. Ser. D* **270**, 2857–2860
Legros, J. J., Franchimont, P. & Hendrick, J. C. (1969) *C.R. Soc. Biol.* **163**, 2773–2777
Lowenstein, W. (1973) *Fed. Proc. Fed. Amer. Soc. Exp. Biol.* **32**, 60–64
Matthews, E. K. (1970) in *Calcium and Cellular Function* (Cuthbert, A. W., ed.), pp. 163–182, St. Martin's Press, New York
Matthews, E. K., Legros, J. J., Grau, J. D., Nordman, J. J. & Dreifuss, J. J. (1973) *Nature (London) New Biol.* **241**, 86–88
McPherson, M. A. & Schofield, J. G. (1972) *FEBS Lett.* **24**, 45–48
Mikiten, T. M. & Douglas, W. W. (1965) *Nature (London)* **207**, 302
Mizel, S. B. & Wilson, L. (1972) *J. Biol. Chem.* **247**, 4102–4105
Mongar, J. L. & Schild, H. O. (1962) *Physiol. Rev.* **42**, 226–270
Nagasawa, J. & Douglas, W. W. (1972) *Brain Res.* **37**, 141–145
Nagasawa, J., Douglas, W. W. & Schulz, R. A. (1970) *Nature (London)* **227**, 407–409
Nagasawa, J., Douglas, W. W. & Schulz, R. A. (1971) *Nature (London)* **232**, 341–342
Nakazato, Y. & Douglas, W. W. (1973) *Proc. Nat. Acad. Sci. U.S.* **70**, 1730–1733
Nicklas, W. J., Puszkin, S. & Berl, S. (1973) *J. Neurochem.* **20**, 109–121
Nordmann, J. J., Dreifuss, J. J. & Legros, J. J. (1971) *Experientia* **27**, 1344–1345
Normann, T. C. (1965) *Z. Zellforsch. Mikrosk. Anat.* **67**, 461–501
Normann, T. C. (1970) *Int. Symp. Neurosecretion 5th* pp. 30–42
Palade, G. (1958) in *Subcellular Particles* (Hayashi, T., ed.), pp. 64–80, Ronald Press, New York
Palay, S. L. (1957) in *Ultrastructure and Cellular Chemistry of Neural Tissue* (Waelsch, H., ed.), pp. 31–44, Hoeber and Harper, New York

Pearse, A. G. E. & Polak, J. M. (1971a) *Histochemie* **27**, 96–102
Pearse, A. G. E. & Polak, J. M. (1971b) *Gut* **12**, 783–788
Poisner, A. M. (1970) in *Biochemistry of Simple Neuronal Models* (Costa, E. & Giacobini, E., eds.), pp. 95–108, Raven Press, New York
Poisner, A. M. & Bernstein, J. (1969) *J. Pharmacol. Exp. Ther.* **177**, 102–108
Poisner, A. M. & Douglas, W. W. (1966) *Proc. Soc. Exp. Biol. Med.* **123**, 62–64
Poisner, A. M. & Douglas, W. W. (1968) *Mol. Pharmacol.* **4**, 531–540
Poisner, A. M. & Hava, M. (1970) *Mol. Pharmacol.* **6**, 407–415
Poisner, A. M. & Trifaró, J. M. (1967) *Mol. Pharmacol.* **3**, 561–571
Pysh, J. J. & Wiley, R. G. (1972) *Science* **176**, 191–193
Röhlich, P., Anderson, P. & Uvnäs, B. (1971) *J. Cell Biol.* **51**, 465–483
Roth, T. F. & Porter, K. R. (1964) *J. Cell Biol.* **20**, 313–332
Rubin, R. P. (1970) *Pharmacol. Rev.* **22**, 389–428
Rubin, R. P., Feinstein, M. B., Jaanus, S. D. & Paimre, M. (1967) *J. Pharmacol. Exp. Ther.* **155**, 463–471
Sachs, H. (1969) *Advan. Enzymol.* **32**, 327–372
Santolaya, R. C., Bridges, T. E. & Lederis, K. (1972) *Z. Zellforsch.* **125**, 277–288
Scharrer, B. & Weitzman, M. (1970) in *Aspects of Neuroendocrinology* (Bargmann, W. & Scharrer, B., eds.), pp. 1–23, Springer, Berlin
Schild, H. (1936) *Quart. J. Exp. Physiol.* **26**, 165–179
Schramm, M. (1968) *Biochim. Biophys. Acta* **165**, 546–549
Smith, A. D. (1971a) *Phil. Trans. Roy. Soc. London Ser. B* **261**, 363–370
Smith, A. D. (1971b) *Phil. Trans. Roy. Soc. London Ser. B* **261**, 423–437
Smith, A. D. (1972) *Biochem. Soc. Symp.* **36**, 103–131
Smith, U. & Smith, D. S. (1966) *J. Cell Sci.* **1**, 59–66
Smith, U., Smith, D. S., Winkler, H. & Ryan, J. W. (1973) *Science* **179**, 79–82
Trifaró, J. M. & Poisner, A. M. (1967) *Mol. Pharmacol.* **3**, 572–580
Trifaró, J. M., Collier, B., Lastowecka, A. & Stern, D. (1972) *Mol. Pharmacol.* **8**, 264–267
Uttenthal, L. O., Livett, B. G. & Hope, D. B. (1971) *Phil. Trans. Roy. Soc. London Ser. B* **216**, 379–380
Uvnäs, B. & Thon, I. (1961) *Exp. Cell Res.* **23**, 45–57
Weitzman, M. (1969) *Z. Zellforsch. Mikrosk. Anat.* **94**, 147–154
Wessels, N. K., Spooner, B. S., Ash, J. F., Bradley, M. P., Luduena, M. A., Taylor, E. L., Wren, J. R. & Yamada, K. M. (1971) *Science* **171**, 135–143
Williamson, J. R., Lacy, P. E. & Grisham, J. W. (1961) *Diabetes* **10**, 460–469
Zigmond, S. H. & Hirsch, J. G. (1972) *Science* **176**, 1432–1434

Calcium Movements in Relation to Contraction

By C. C. ASHLEY and P. C. CALDWELL

*Department of Physiology and Department of Zoology,
University of Bristol, Bristol BS8 1UG, U.K.*

Synopsis

Calcium changes within intact single muscle fibres have been investigated by employing radioactive calcium as well as the calcium-sensitive protein aequorin. The efflux of ^{45}Ca from the cell after injection was not initially exponential but it became so after a period of 1–2h. If the fibre was maximally stimulated at this stage, either electrically or by the application of potassium or caffeine in physiological saline, the fibre contracted and the calcium efflux increased some 10–15-fold compared with resting values. Pretreatment of the fibre with lower concentrations of these agents produced smaller responses and virtually abolished the ability of the fibre to respond to subsequent maximal activation. Recently an optical technique using the photoprotein aequorin has enabled a better resolution of the time-course of the changes in free calcium to be achieved. A detailed analysis of the time-course and magnitude of the calcium changes in relation to the isometric tension response has indicated that a kinetic scheme in which two calciums are each involved in a relatively slowly equilibrating reaction in the same functional unit makes accurate predictions as to the nature of recorded tension. This scheme for the co-operative action of two calciums in the same functional unit also makes accurate predictions as to the relationship between steady-state adenosine triphosphatase and calcium bound in isolated myofibrils. On this basis the contribution of the sarcoplasmic reticulum to the variation in free calcium can be deduced from the experimental records, and the results have suggested mechanisms by which calcium release and reaccumulation by the sarcoplasmic reticulum may occur within intact muscle fibres.

Introduction

It has been known for a considerable number of years that calcium ions play an important part in contraction. The early observations of Ringer (1883) on cardiac muscle were probably the first to implicate a crucial role for calcium, but it was not until the work of Heilbrunn and his associates, some 60 years later, that calcium ions were shown to be important in skeletal as well as cardiac muscle (Heilbrunn & Wierciński, 1947). In these experiments a number of physiologically occurring ions were injected directly into the sarcoplasm of isolated frog muscle fibres; only those of calcium were shown to be effective in initiating a contraction. Heilbrunn suggested that calcium was released from the outer regions of the muscle fibre and passed inwards by diffusion, activating the contractile machinery in the process. It was impossible, however, to explain the rapid contraction of vertebrate skeletal muscle on the basis of the simple diffusion of calcium from the outer regions of the muscle fibre to the centre (Hill, 1948, 1949). If calcium ions were to be directly involved in contraction, either some process other than simple diffusion must be involved or else calcium must be stored at sites within the muscle fibre considerably closer to the contractile machinery (see Sandow, 1952). In more recent years it has been well established that the longitudinal elements of the sarcoplasmic reticulum, the network of tubules which form a 'collar' around each myofibril containing the contractile proteins, provide the sites for the release and reaccumulation of calcium during a contractile response. Thus the diffusion distance for calcium is considerably decreased in vertebrate skeletal muscle to the order of $1-2\,\mu m$. On this basis, the site of release of calcium from the sarcoplasmic reticulum is still, however, spatially separated from its site of action on the functional units of the myofibril, which in the majority of species contain the calcium-binding complex, troponin (Ebashi et al., 1969; Lehman et al., 1972; Perry et al., 1972; Drabikowski et al., 1972; Greaser et al., 1972). It is therefore the movement of calcium from the site of release to its site of action, through the fibre sarcoplasm, that must be detected if the kinetic relationship between the free calcium concentration and force development are to be elucidated.

As single vertebrate muscle fibres are fairly small, $10-150\,\mu m$ in diameter, it is often convenient to use larger fibres for physiological investigations as these are much more readily dissected and manipulated. There are a number of invertebrate species that have large single muscle fibres, particularly in the Crustacea. The spider crab *Maia squinado* and the Pacific barnacle *Balanus nubilus* have single striated muscle fibres that are 1–2 mm in diameter and 2–3 cm in length (Caldwell, 1958; Caldwell & Walster, 1963; Hoyle & Smyth, 1963). The fibres once isolated can be cannulated and this manipulation permits materials to be injected in micro amounts directly into the interior of the muscle fibre without additional damage to the outer membrane. The procedure of injection of micro amounts of material seems to produce no serious damage within the cell either to the contractile proteins or to the elements of the sarcoplasmic reticulum. There are a number of techniques available by which the internal movements of calcium can be detected in conjunction with these large single muscle fibres. If, however, a quantitative relationship between the free calcium concentration and force development is to be examined, it is essential that both the time-course and the magnitude of these calcium movements be known with some degree of precision. A sensitive method that has provided some information about intracellular calcium changes is the use of ^{45}Ca.

Use of $^{45}CaCl_2$ to Detect Internal Calcium Movements

The method depends on the fact that there is always an efflux of calcium across the outer cell membrane associated with major calcium movements through the fibre sarcoplasm. It is possible to follow the time-course and magnitude of this calcium efflux from the fibre and to use these parameters as indices of intracellular calcium changes. For the technique to be feasible, the internal store of calcium within the sarcoplasmic reticulum must be initially labelled with ^{45}Ca. This can be achieved very readily in large crustacean muscle fibres by the direct technique of injection of micro amounts of materials, mentioned above.

After the injection, the ^{45}Ca loss from the fibre is measured and expressed as a rate constant (k), which is given by (counts lost/min during the collection period)/(average counts remaining in fibre during collection period). If the ^{45}Ca solution injected results in more than approx. 0.1 μM of free calcium ion then the initial calcium efflux does not follow first-order kinetics until after a period of 1–2 h (Fig. 1). If, however, the ^{45}Ca is injected in combination with the chelating agent EGTA [ethanedioxybis(ethylamine)tetra-acetic acid], so that the concentration of free calcium ion injected is less than approx. 0.1 μM, then the loss does initially approximate to first-order kinetics and the rate constant, after the initial period of radial diffusion is complete, is roughly constant with time (Fig. 2). The kinetic behaviour of the calcium efflux, depending upon the free calcium concentration injected, suggests that at high calcium concentrations there is a considerable removal from the sarcoplasm of free calcium into a compartment from which it is lost only slowly with time. It is possible to estimate from the counts of radioactivity lost from the fibre during the early stages of rapid calcium efflux that, of the total calcium injected, some 60% is retained by the fibre, whereas the remaining 40% is lost in the first 1–2 h by efflux across the outer membranes. If the concentration of calcium injected is greater than approx. 0.1 μM, the fibre contracts, but this is followed by relaxation within 1–3 min of the injection. This observation implies that within 1–3 min the internal free calcium concentration has returned to close to its resting value. It is possible to estimate the value of the resting calcium concentration in these large muscle fibres by two separate methods. The first is to employ calcium buffer solutions containing different concentrations of stabilized free calcium ion. These are injected into the interior of the fibres and

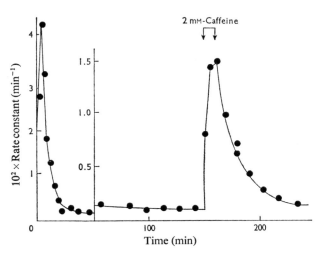

Fig. 1. *Effect of 2 mM-caffeine in saline upon the residual calcium efflux from a single Maia muscle fibre*

The composition of crab 'saline' used throughout the present paper is standard (Fatt & Katz, 1953). The efflux increases by some 10–15 times and this is accompanied by a strong contraction (from Ashley et al., 1972). The y-axis scale was adjusted at time indicated.

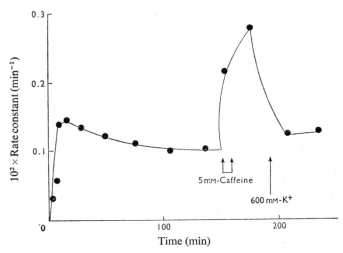

Fig. 2. *Effect of injecting* $^{45}CaCl_2$ *in conjunction with the complexing agent EGTA (free* $[Ca^{2+}]$ *approx. 20 nM)*

The initial rapid phase of loss is absent. Application of 5 mM-caffeine in saline stimulates the efflux by two to three times. Subsequent application of 600 mM-K^+ in saline was ineffective. Single *Maia* muscle fibre (from Ashley *et al.*, 1972).

the concentration of free calcium noted at which a just-detectable contraction is elicited. These investigations indicated that calcium concentrations greater than approx. 0.3 μM caused a contraction and this suggested that the resting calcium concentration in the fibre sarcoplasm should be less than approx. 0.3 μM (Portzehl *et al.*, 1964; Ashley, 1967). A second method, with the calcium-sensitive photoprotein aequorin (Shimomura *et al.*, 1962, 1963; Shimomura & Johnson, 1969), indicates that the resting calcium concentration in these large fibres is in the range 0.07–0.1 μM at 10–12°C (Ashley, 1970, 1971). On this basis, it is now possible to estimate the efflux of calcium from these large fibres during the initial non-exponential phase of the efflux, bearing in mind the value for the resting sarcoplasmic calcium concentration. The rate of pumping calcium from the fibre, not taking into account surface area of the invaginations of the outer membrane (the cleft system; see below) but assuming that the transport system does not saturate, must be approx. 8 pmol/s per cm^2 in order to decrease the internal free calcium after injection to approx. 0.1μM within 2 min. This value is not sufficiently different from the residual efflux rate observed after 2 h to warrant the postulation of an additional more specialized calcium-transport system across the outer cell membrane operating during the early stages of the efflux. The residual rate of calcium efflux is approx. 0.4–2 pmol/s per cm^2 at 15–22°C for both *Maia* and *Balanus* (Ashley *et al.*, 1972) and the mechanism seems to be a sodium–calcium and calcium–calcium exchange system across the cell membrane of both of these large-muscle-fibre preparations (Ashley & Ellory, 1972*a*, 1973; Ashley *et al.*, 1973*a*; Baker, 1972).

The fraction of calcium that is lost completely from the fibre must also be removed from the sarcoplasm within the 1–3 min period. Nevertheless, the initial efflux takes of the order of 1–2 h to be decreased to the residual value characteristic of the resting state of the fibre. The explanation of this time-lag may depend on the complex morphology of these muscle fibres. There are numerous invaginations of the outer membrane (the cleft system) and these form narrow tortuous channels whose lumina are in direct contact with the external saline and yet which penetrate deeply into the fibre cross-section. The surface area of these invaginations has been estimated as 15–20 times the simple surface area of the fibre if it is considered to be a cylinder (Selverston, 1967). This complex tubular system may delay considerably the calcium efflux as a function of time, compared with the time-course of the free calcium change within the fibre. This is true for both the efflux observed after the initial injection and also for that resulting from internal calcium changes produced by contractile agents applied to the fibre.

It is likely that the 60% of the injected calcium that remains within the fibre is removed into the sarcoplasmic reticulum compartment. In crustacean muscle, the elements of the sarcoplasmic reticulum have a marked ability to accumulate calcium, compared even with vertebrate skeletal muscle, and values of approx. 1.7 μmol/min per mg of sarcoplasmic reticulum protein have been reported by Baskin (1971) both for lobster and for barnacle. The mitochondrial content of these crustacean fibres is low (Hoyle et al., 1972). These are generally not located between the myofibrils, but are just below the outer membrane. It seems from both the position of the mitochondria and from their sparcity, that they can play little part in calcium movements associated with contraction in these fibres. Also an investigation into the metabolism of these large muscle fibres suggests that mitochondria play only a small role in ATP production, the metabolism relying mainly on anaerobic glycolysis. Stimulation of the fibre either electrically or by the application of saline solutions containing a high concentration of potassium ions or caffeine produced a mobilization of calcium from the sarcoplasmic reticulum. This process is associated with an increased rate of loss of ^{45}Ca across the outer membrane and is illustrated in Fig. 1 for the application of 2 mM-caffeine in saline to a single *Maia* muscle fibre. [The composition of crab 'saline' used throughout the present paper is standard (Fatt & Katz, 1953).] The rate constant for the efflux is increased 10–15 times by caffeine, which at this concentration would be expected to activate fully the sarcoplasmic reticulum calcium-release mechanism. Less intense amounts of stimulation release less calcium from the sarcoplasmic reticulum, hence the rate constant for the loss of calcium across the outer membrane is increased to a lesser extent, when compared with the residual value. In *Maia* fibres the rate constant for the efflux of calcium from the fibre appears to be linearly related to the magnitude of the free calcium change within the muscle fibre (Ashley et al., 1972). This conclusion is based upon parallel experiments with the intracellular calcium indicator aequorin, which indicate that the free calcium increases some 10–15 times over the resting value with 2–5 mM-caffeine in saline present externally. Thus the use of radioactively labelled calcium can give, under certain conditions, an indirect estimate of the magnitude of the internal calcium changes, but these changes are considerably delayed compared with the true time-course of the free calcium change in the sarcoplasm.

The use of ^{45}Ca as an internal indicator of calcium movements can also give some insight into the mechanisms that may control the normal processes of calcium release and reaccumulation *in vivo*. It is observed, for example, that pre-

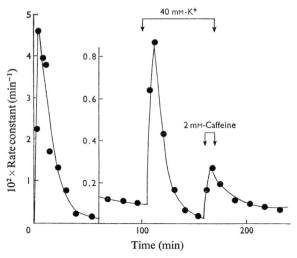

Fig. 3. *Effect of subcontractile concentrations of potassium (40 mm-K^+) in saline followed by 2 mm-caffeine in 40 mm-K^+-containing saline on the residual calcium efflux from a single Maia muscle fibre (from Ashley et al., 1972)*

For details see the text. The *y*-axis scale was adjusted at the time indicated.

treatment of the muscle fibre with saline solutions containing increased amounts of potassium or low concentrations of caffeine, both of which are themselves insufficient to initiate a visible contraction, produced a detectable increase in the efflux of calcium from the fibre over the resting value. This is illustrated for a challenge by a subthreshold amount (40 mM) of potassium in Fig. 3. If the fibre is now challenged by a concentration of caffeine or raised amount of potassium that would normally be expected to initiate full activation of the contractile system together with a 10–15 times increase in the rate constant for calcium efflux, little or no contractile response or increase in efflux is in fact obtained. It seems that pretreatment with subthreshold contractile agents also mobilizes calcium from the sarcoplasmic reticulum, but that if this process is prolonged, either a depletion of the releasable calcium fraction occurs, or else the process of calcium release fatigues so that higher concentrations of these agents are now ineffective (Fig. 3, cf. Fig. 1).

If the calcium is injected as a complex with the chelating agent EGTA, so that the free calcium concentration was small (less than $0.1 \mu M$), there was, as already noted, no initial rapid phase of calcium loss. The efflux after radial diffusion was complete (15–20 min) was approximately exponential. Similarly, if EGTA alone was injected into a previously ^{45}Ca-loaded muscle fibre, at the stage of low residual efflux, the loss of calcium was little affected, decreasing by only a factor of two to three times and then gradually increasing to the original value (see Fig. 4 and below). If, however, contractile agents (e.g. 2 mM-caffeine in saline) are applied to these fibres in which EGTA is present internally, both the fibre contraction and the stimulated calcium efflux are considerably smaller (two to three times the residual efflux) than that observed in the absence of EGTA internally (Fig. 2, cf. Fig. 1) (Caldwell, 1970). There are a number of conclusions that can be drawn

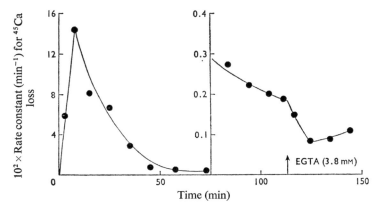

Fig. 4. *Effect of the injection of EGTA (↑) on the residual calcium efflux from a single Balanus muscle fibre*

The final EGTA concentration after fibre dilution was approx. 3.8 mmol/kg wet wt. (from Ashley et al., 1972). The second y-axis scale is an expansion of part of the first one.

from these experiments. The observation that internal injections of EGTA are able virtually to suppress contraction, strongly suggests that the major fraction of the calcium normally required for contraction is derived from stores within the longitudinal elements of the sarcoplasmic reticulum, rather than from an external source. This conclusion is supported by the fact that the concentration of EGTA required is very similar to the total fibre calcium concentration (Ashley, 1967). The second point is that the EGTA injection experiments produce only a small change in the free calcium concentration, as judged by the change in the efflux of ^{45}Ca, and also from the change in aequorin light emission (Fig. 5). This suggests

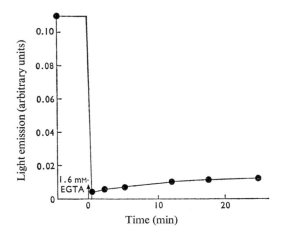

Fig. 5. *Effect of injecting EGTA (↑) on the resting rate of light emission from an aequorin-injected single Balanus muscle fibre*

The final EGTA concentration was estimated after fibre dilution to be approx. 1.6 mmol/kg wet wt. The relative change in free calcium concentration is given by the square root of the ratio of the initial and final light emission.

that the calcium stores within these muscle fibres must be readily releasable, so as to maintain the sarcoplasmic calcium concentration at close to its resting value of 0.07–0.1 μM despite the EGTA injection. If the calcium stores, e.g. the sarcoplasmic reticulum, were either not readily releasable or not turning over at a relatively rapid rate, then calculations suggest that the free calcium concentration should be lowered as a result of the injection of EGTA by a factor of some 100 times, to the region of 1 nM, and this seems to be the case in squid axon. Here the rate constant for the loss of ^{45}Ca from the fibre is decreased by some 100 times by the injection of millimolar concentrations of EGTA (Blaustein & Hodgkin, 1969), suggesting that the free calcium in the axoplasm has been decreased by a similar extent from its resting value of approx. 0.1 μM (Baker et al., 1971). Therefore experiments with radioactively labelled calcium can, in these preparations, give a considerable amount of information about the magnitude of the internal calcium changes, together with some insight into the processes involved in excitation–contraction coupling, that is the series of steps by which depolarization of the outer cell membrane leads to the release of calcium and the development of tension. It does not provide, however, at least in the experiments reported here, any precise information about the actual time-course of calcium movements within the fibre sarcoplasm during contraction and relaxation.

Optical Measurements of the Free Calcium Change

The calcium indicator, murexide, has frequently been used to follow calcium changes *in vitro* as a function of time. In particular, the rate at which vesicles isolated from elements of the sarcoplasmic reticulum accumulate calcium has been investigated by using conventional double-beam spectrophotometric methods with murexide in the supporting medium (Ohnishi & Ebashi, 1963, 1964; Harigaya et al., 1968; Harigaya & Schwartz, 1969; Inesi & Scarpa, 1972). The use of murexide within intact muscle cells is complicated by the fact that associated with each mechanical response there is also a large change in light scattering. The small absorption change due to the formation of the calcium–murexide complex ($K_{Ca-murexide}$, approx. $1 \times 10^3 \, M^{-1}$) has therefore to be distinguished from the much larger change resulting from contraction. It is possible to measure this small absorption change by using single-beam spectrophotometric methods in conjunction with averaging techniques (Jöbsis & O'Connor, 1966). In toad muscle that has been allowed to accumulate murexide there is a relatively rapid formation of the calcium–murexide complex which precedes the development of tension and returns to its initial value before maximum tension (Fig. 6). It was not possible for technical reasons to follow the time-course of the formation of the calcium complex during relaxation. These experiments suggest that in amphibian muscle the free calcium concentration, which is likely to determine active state, rises relatively slowly and can be graded in intensity, since more complex was formed during a tetanus than during a single response (Jöbsis & O'Connor, 1966). However, the difficulties associated with examining the time-course of calcium release during relaxation, together with technical problems with the averaging method on single fibres, suggested that other techniques should be investigated. The most promising seemed to be that based on the use of the calcium-sensitive

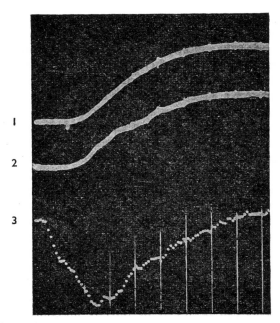

Fig. 6. *Time-relations between tension development (trace* 1), *light-scattering (trace* 2) *and the kinetics of the calcium–murexide complex (trace* 3)
Time, 25 ms per division. The stimulus was delayed 12 ms after the start of the sweep. Temperature, 12°C. Toad sartorius muscle was used (from Jöbsis & O'Connor, 1966).

photoprotein aequorin (Shimomura *et al.*, 1962, 1963; Shimomura & Johnson, 1969, 1970).

Use of the Photoprotein Aequorin as an Intracellular Calcium Indicator

Perhaps the major advantage of aequorin compared with murexide is that the aequorin method is a luminescence method independent of an external energy input, whereas the murexide method is an absorption method which is implicitly more complex. In the isolated state, aequorin is a precharged molecule (mol.wt. approx. 30000) which requires neither ATP nor O_2 for the luminescence reaction. The light-emission curve has a maximum at about 470 nm and the photon emission is evidently the result of an intramolecular rearrangement with the release of more than 60 kcal·mol^{-1}. The isolated protein, which is extracted from the hydromedusa *Aequorea forskalea* (Shimomura & Johnson, 1969), does not act enzymically with calcium. The product protein formed as a result of the luminescence reaction is strongly fluorescent, the action spectrum having a similar maximum to the luminescence spectrum of the native protein. In addition, the fluorescent protein is still able to bind calcium reversibly and this process considerably enhances the fluorescence maximum (Shimomura & Johnson, 1969, 1970). Other ions, besides those of calcium, are able to initiate the luminescence reaction, although they are not likely to be of physiological importance (Shimomura *et al.*, 1963; Izutsu *et al.*, 1972). Fortunately, magnesium ions which are at a relatively high concentration in the muscle cells (Ashley & Ellory, 1972*b*) are not effective in the light-emitting reaction, although they do compete with calcium for the active site (see later).

In order for aequorin to be a successful indicator for intracellular calcium movements, it must satisfy a number of criteria. First, it must be responsive to free calcium in the range of concentrations encountered physiologically, that is 0.1–10 μM. Therefore for a large luminescence change *in vivo* the photoprotein should have sites with a binding constant for calcium similar to the reciprocal of the range of free calcium concentrations encountered *in vivo*, that is 1×10^5–$1 \times 10^7 \text{M}^{-1}$. The concentration of the indicator must not interfere with the kinetic interaction between calcium in the sarcoplasm and the calcium bound to the sites on the myofibrils. It would also be a distinct advantage if the luminescence change could be 'amplified' by a relationship involving more than 1 calcium for light emission. Nevertheless, it is of importance that both the precise nature of the relationship between free calcium and light emission, as well as the time-response of the aequorin system to rapid changes in free calcium, can be readily determined. Finally, the protein must be tolerated by the physiological preparation. It must not have a toxic effect upon the processes of calcium release or reaccumulation by the sarcoplasmic reticulum, nor must it have an inhibitory effect upon the myofibrillar ATPase (adenosine triphosphatase). The first and last of these requirements can be tested by directly injecting the photoprotein into large single muscle fibres. The tension responses can then be examined from this preparation, as a result of a given depolarization, over a period of several hours. In these experiments, the protein was prepared from the medusal extract by conventional methods (Shimomura & Johnson, 1969) and a solution of the purified protein in iso-osmotic KCl injected axially into the muscle fibre. A period of 1–2h was allowed for radial diffusion and the final protein concentration was estimated, assuming uniform distribution, as about 1μM. At rest, there was a characteristically low rate of light emission from the muscle fibre, but this rate was considerably increased upon stimulation (Ridgway & Ashley, 1967; Ashley & Ridgway, 1968, 1970). Examination of the fibre over a period of several hours indicated that, in comparison with control fibres not injected with aequorin, the photoprotein had no significant effect upon the fibre responses. It seemed that aequorin had no adverse effect upon either the processes of excitation–contraction coupling or upon the myofibrillar tension-generating apparatus. Estimates of the amount of aequorin utilized per impulse suggested that it was less than 1 nM; the main consumption of the photoprotein was in the resting state when considered over the time-course of an experiment lasting 4–5h, despite the low calcium concentration and consequent low rate of resting light emission. These initial experiments revealed that there was no simple relationship between free calcium concentration and the end-product, that is the tension developed by the muscle. It is worth pointing out that in these fibres, the electrical event upon the outer membrane is usually not an all-or-nothing process as in most vertebrates and the intensity of depolarization and hence calcium release can be uniformly graded by using a wire electrode inserted axially into the fibre and different intensity impulses. The calcium concentration in the sarcoplasm, as judged by the change in aequorin light emission, increased soon after membrane depolarization and preceded the development of tension. The light emission returned virtually to the resting value at peak tension and there was no increase in light emission during the phase of relaxation (Fig. 7). At short stimulus durations, the time-course of the aequorin light emission (calcium-transient) was biphasic, the precise extent of each phase being dependent upon the intensity and duration of the membrane depolarization (Fig. 7a). As a result of longer stimulus durations or brief tetani, the calcium transient showed a more complicated relationship with time (Fig. 7b and 7c), but here again the phases were governed by the intensity and duration of the stimulus (Ashley & Ridgway, 1968, 1970). It is possible to derive a considerable amount of information concerning the mechanisms by which changes in membrane potential control the process of calcium release from the sarcoplasmic reticulum on the basis of this type of experimental record. However, as a first stage in the analysis, it is important to determine the way that changes in the free calcium concentration influence the development of tension. Preliminary analysis of the relationship suggested that the free calcium concentration controlled the rate of development of tension (Ridgway & Ashley, 1967; Ashley & Ridgway, 1970), so that the peak of the light emission occurred at a time of maximum rate of development of force. It seemed that the reactions in which calcium ions were involved to produce tension, equilibrated relatively slowly with time (Ashley & Moisescu, 1972b).

In order to convert this qualitative observation into a quantitative relationship, the precise size and time-course of the free calcium change must be known from the aequorin response. To determine the magnitude of the calcium changes indicated by the light emission, a number of different techniques were used, all of which suggested that there was a square-law relationship between free calcium

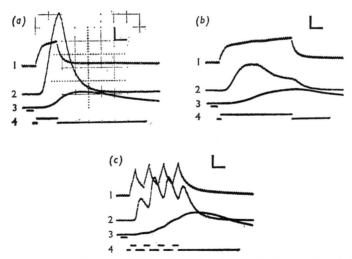

Fig. 7. *Effect of (a) short, (b) long and (c) brief tetanic pulses on the time-relations between membrane potential (trace 1), aequorin light emission (calcium-transient)(trace 2) and isometric tension (trace 3)*

The calibration and stimulus marks are on trace 4. Nominal intensity, 3.0. Duration, (a) 150, (b) 500 and (c) 40ms. Calibration; bar, vertical: trace 1, 10mV; trace 2, (a) and (c) 1.9nlm; (b) 3.8nlm; trace 3, 2.5g wt; horizontal; 100ms. Temperature, 10–12°C. All traces are from the same single *Balanus* muscle fibre. Resting light emission: (a) 0.6nlm, (b) 1 nlm and (c) 0.27nlm (from Ashley & Ridgway, 1970).

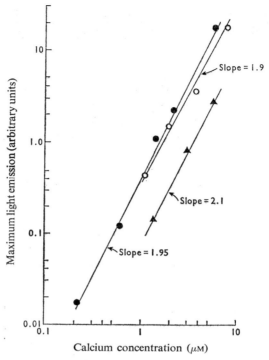

Fig. 8. *Plot of the maximum rate of light emission from a mixing chamber in vitro against free calcium concentration; both variables are plotted on a logarithmic scale*

Chamber volume, 50 μl. ●, ○, 5 mM-magnesium; ▲, 10 mM-magnesium. ○, Unbuffered calcium solution, pH 6.8–7.5, slope 1.9; ●, calcium–EGTA buffer solution, pH 7.1, slope 1.95; ▲, calcium–EGTA buffer solution, pH 7.1, slope 2.1 (from Ashley, 1970). Temperature, approx. 20 °C. Aequorin concentration was approx. 0.3 μM. KCl (50 mM) was present. Assumed apparent binding constant for calcium to EGTA = 7.6×10^6 M^{-1}, pH 7.1.

and aequorin light emission, so that a doubling of the calcium concentration produces a quadrupling of the rate of light output (Ashley, 1970, 1971; Baker *et al.*, 1971). The most precise method to determine this relationship was a mixing chamber *in vitro*, similar in size to the intact muscle fibre. This technique allowed the effects of various aequorin and magnesium concentrations, as well as the ionic strength, on the relationship between calcium concentration and light emission to be measured. In addition it permitted a rough estimate to be made of the values of resting fibre light output in terms of absolute calcium concentration, since the geometry of the chamber with respect to the face-plate of the photomultiplier tube was virtually the same as that of the individual muscle fibre. The results from these investigations are summarized in Fig. 8, where the log of the maximum light emission obtained after mixing extrapolated to zero time, is plotted against the log of the calcium concentration. A straight-line relationship, with a slope of 2 is maintained for free calcium concentrations up to about 10 μM, at least in the presence of the two different free magnesium concentrations, illustrated. The values of magnesium concentration investigated in Fig. 8 appear to be in the range of diffusible magnesium encountered in these single muscle fibres (Ashley & Ellory,

1972b). These observations suggest that the apparent value of $K_{Ca-aeq.}$ must be approx. $1 \times 10^5 M^{-1}$ per site in the muscle and this finding has additional support from calculations on the basis of the rate of consumption of aequorin *in vivo* (Ashley & Moisescu, 1972b). In the absence of magnesium, the square-law relationship operates over a lower range of calcium concentrations and the relationship deviates from a square law as the free calcium concentration approaches the value of the reciprocal of $K_{Ca-aeq.}$, which is now greater than $1 \times 10^6 M^{-1}$ per site (C. C. Ashley, D. G. Moisescu & R. M. Rose, unpublished work). Thus the major effect of increasing the magnesium concentration is to shift the relationship along the calcium concentration axis, so that increasing the magnesium concentration decreases the apparent binding constant for calcium (van Leeuwen & Blinks, 1969; Hastings *et al.*, 1969; Ashley, 1970; Baker *et al.*, 1971). The square-law relationship between calcium and aequorin light emission implies that two calcium ions are required per active site on the photoprotein, although the precise kinetic mechanism by which this occurs has not been investigated in any detail.

Similarly, a square-law relationship can also be demonstrated if the muscle fibres are used as mixing chambers. The fibres are first loaded with aequorin and then injected with excess of calcium buffer solution of known free calcium concentration (Ashley, 1970). In addition, the injection experiments on intact muscle fibres indicate that the resting calcium concentration is in the range of 0.07–0.1 μM at approx. 10°C (Ashley, 1970, 1971). Finally, the results from two different techniques as to the effect of the injections of EGTA on the resting calcium concentration in the sarcoplasm also suggest that the relationship between light emission and calcium concentration *in vivo* is a square-law. These are illustrated in Figs. 4 and 5 where the effect of injecting EGTA on the efflux of ^{45}Ca from the fibre is examined, and in a separate experiment on the resting light emission from an aequorin-loaded muscle fibre. The result on the ^{45}Ca efflux has already been discussed and suggests that the resting calcium concentration decreases only by a factor of two to three times and then gradually increases towards the initial value. The effect on the aequorin emission is to decrease the resting rate by some 10 times but this gradually increases over a period of 10–20 min towards the original value (Fig. 5). Thus for a square-law relationship, the light-emission change suggests a decrease in the resting calcium concentration of about three times, which is close to the value indicated from the efflux experiment.

The size of the calcium change during electrical stimulation is illustrated for a typical fibre response in Fig. 9, on the basis of the calibration of the aequorin light emission. Here a low-intensity depolarization produces, roughly speaking, a threefold change in the free calcium concentration and results in a tension response which is about 10% of maximum obtainable tension (P_0) (Ashley, 1971).

In addition to the knowledge of the size of the calcium change associated with tension development, it is also of importance to know the time-course of the changes with some degree of precision, if a quantitative relationship between free calcium and tension is to be explored. Rapid-mixing techniques with a stopped-flow apparatus (dead time 1–2 ms) have been used to investigate the kinetics of the calcium–aequorin (aeq.) reaction *in vitro* (Hastings *et al.*, 1969; van Leeuwen & Blinks, 1969). The scheme proposed, consistent with the experimental observations, is indicated below and has been adapted for 2 calcium ions. The first step is

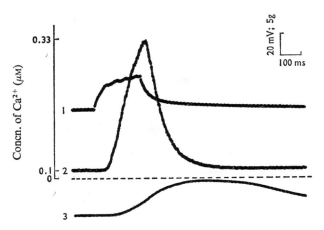

Fig. 9. *Time-relations between membrane depolarization (trace 1), aequorin light emission (trace 2) and isometric tension (trace 3)*
Record was from a single *Balanus* muscle fibre (from Ashley, 1971). Temperature, 10–12°C.

a rapid, reversible binding of the two calcium ions and the formation of the component of two Ca–aeq.

$$2Ca^{2+} + aeq. \underset{k'_{-}}{\overset{k'}{\rightleftharpoons}} 2Ca\text{–aeq.} \overset{k''}{\longrightarrow} X \overset{k'''}{\longrightarrow} Y^* \overset{Fast}{\longrightarrow} Y + h\nu$$

where k' and k'_{-} are assumed the same for both sites. When rapidly mixed (2–3ms) the pre-steady state and hence the rise time of the light emission is governed mainly by k''', which has an interpolated value of $80\,s^{-1}$ at 10°C, the temperature of the muscle-fibre experiments. When aequorin is saturated the decay of light is dependent upon k'', which has a value of about $1\,s^{-1}$ at 20°C (Hastings et al., 1969; van Leeuwen & Blinks, 1969) and now the rate of increase of the light emission was virtually unaffected by free calcium concentration (Hastings et al., 1969). The light-emission traces recorded from the stimulated muscle fibre can be readily corrected on the basis of this kinetic scheme. Essentially, the correction results in a free calcium transient which is translated, with virtually the same shape, some 12ms in front of the aequorin light-emission trace (Ashley & Moisescu, 1972a,b). The correction is small since the rise time of the aequorin light emission is relatively rapid compared with the rate of rise of the free calcium concentration in the sarcoplasm. A scheme proposed more recently to explain the light emission from aequorin, on the basis of measurements made with a rapid-mixing continuous-flow apparatus (Loschen & Chance, 1971), results in virtually the same conclusions as to the time-course of the free calcium event in the muscle-fibre experiments.

Kinetic Model for the Action of Calcium

It is important for the development of a kinetic relationship between free calcium and tension to know both the number and the stoicheiometry of the controlling reactions in which calcium is involved. The preliminary observations on the relationship between free calcium and the rate of development of tension, implied that at least one slow calcium-dependent reaction was important in tension development. That is to say in the early stages in the development of tension, providing that calcium is considered to be involved in reversible reactions, the forward reaction step is relatively more important than the reverse, so that the product tension lags behind the changes in free calcium. Analysis of the kinetic experiments with aequorin *in vivo* (Ashley & Ridgway, 1970), indicates that two calcium ions and two relatively slow reactions are involved per functional unit for tension production (Ashley & Moisescu, 1972a,b). There are three basic kinetic schemes by which two calcium ions can be combined with two relatively slow reactions to give rise to tension. The two schemes that provide the best reconstruction of the recorded tension event, on the basis solely of the value of the free calcium concentration at any instant in time, involve the co-operative binding of two calcium ions in the same functional unit. They both provide a good fit to the recorded tension response under conditions when they are mathematically identical. In one case, the independent scheme, the two sites (M_1, M_2) in the functional unit are considered to be independent in their calcium-binding ability, but only when both are occupied with calcium can the component ($CaM_1 \infty M_2Ca$), which is proportional to the product of the concentrations of CaM_1 and CaM_2, give rise to tension. This scheme can be expressed as follows:

$$M_1^* \infty M_2^* + Ca^{2+} \underset{}{\overset{Slow}{\rightleftharpoons}} \begin{Bmatrix} CaM_1 \infty M_2^* \\ \text{or} \\ CaM_2 \infty M_1^* \end{Bmatrix} + Ca^{2+} \underset{}{\overset{Slow}{\rightleftharpoons}} CaM_1 \infty M_2Ca \sim \text{tension}$$
(ATPase)

where an asterisk represents an active calcium-binding site. The concentration of $M_1^* \infty M_2^* + CaM_1 \infty M_2^* + CaM_2 \infty M_1^* + CaM_1 \infty M_2Ca$ is constant and is the concentration of the functional unit. In the alternative scheme, a consecutive reaction, the second calcium ion can only be bound after the first and this can be indicated as follows:

$$M_1^* \infty M_2 + Ca^{2+} \underset{}{\overset{Slow}{\rightleftharpoons}} CaM_1 \infty M_2^* + Ca^{2+} \underset{}{\overset{Slow}{\rightleftharpoons}} CaM_1 \infty M_2Ca \sim \text{tension}$$
(ATPase)

where concentration of $M_1^* \infty M_2 + CaM_1 \infty M_2^* + CaM_1 \infty M_2Ca$ is constant and is the concentration of the functional unit.

If the independent scheme is developed in more detail for the case where the calcium concentration is varying with time (t), as is the situation with the muscle-fibre experiments, then for a site i, where $i = 1,2$ in this case,

$$Ca^{2+} + M_i \underset{k_{-i}}{\overset{k_i}{\rightleftharpoons}} CaM_i$$

and where $M_i + CaM_i = c_i$ and c_i is the concentration of the calcium-binding sites that lead to tension production, then it can be shown that:

$$CaM_i(t) = c_i \bigg(1 - \exp\{k_{-i}[-K_i\,Ca_r\,I(t) - t]\} \bigg\langle k_{-i}\!\int_0^t \exp\{(k_{-i}[K_i\,Ca_r\,I(t') + t']\}\mathrm{d}t'$$
$$+ (1 + K_i\,Ca_r)^{-1} \bigg\rangle \bigg) \qquad (1)$$

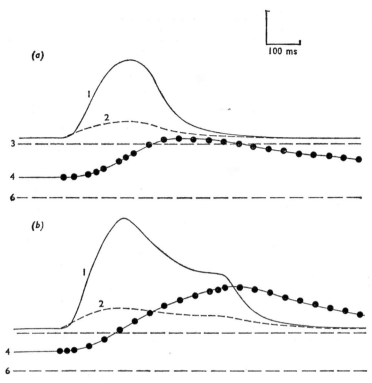

Fig. 10. *Effect of short- and long-duration stimuli upon the time-relations between aequorin light emission (trace 1), calculated free calcium from aequorin kinetics ($k''' = 80\,s^{-1}$ at $10°C$) (trace 2), baseline for light and free calcium (trace 3), isometric tension (trace 4),* ●, *tension calculated from eqns. (1) and (2), baseline for tension and calcium bound (trace 6)*

$P_{max.}$ maximum tension per response for (a) is 5.6%, and for (b) is 8.3% P_0 (maximum obtainable tension). Temperature, 10–12°C. The same normalization of ±4% was used for (a) and (b) (from Ashley & Moisescu, 1972b). Tracings were from the same isolated *Balanus* muscle fibre. Calibration bar; vertical: trace 1, 3.8 nlm; trace 2, $Ca_r \times 6.67$; trace 4, 2.5 g wt. (2.6% of P_0).

where $K_i = k_i/k_{-i}$, $I(t) = \int_0^t Ca(t')/Ca_r \, dt'$ and Ca_r is the resting fibre calcium concentration. Therefore from the model, tension is given as follows:

$$\text{Tension (ATPase)} \sim \prod_i CaM_i \qquad (2)$$

and

$$\text{Calcium bound} \sim \sum_i CaM_i \qquad (3)$$

The results from eqns. (1) and (2) are indicated in Fig. 10, where the value of Ca/Ca_r was determined from the experimentally recorded aequorin light emission. The value of $K_i Ca_r$ was taken as 0.15, as the resting calcium concentration in these fibres is in the range 0.07–0.1 μM (Ashley, 1970) and K_i can be calculated, from steady-state tension versus calcium concentration relationship for barnacle, as between 1×10^6 and $2 \times 10^6 \, M^{-1}$ per site, depending upon the free magnesium concentration (Ashley & Moisescu, 1973b; C. C. Ashley, D. G. Moisescu &

R. M. Rose, unpublished work). The value of k_{-t} estimated from the falling phase of tension as about $1 \, s^{-1}$, was taken as $1.29 \, s^{-1}$ at 10 °C. Other details are mentioned in the legend (see also Ashley & Moisescu, 1972b). Calculations, in which the process of diffusion from the site of release to the site of uptake is also considered, result in virtually the same conclusions for predicted tension (Ashley et al., 1973b).

The important fact that emerges from these calculations is that it is possible to determine very precisely the value of relative isometric tension as a function of time simply from a knowledge of the free calcium concentration at any instant. The functional unit, which is involved in the co-operative binding of the two calcium ions, must presumably involve the protein troponin which is located on the actin filaments in these fibres (Lehman et al., 1973). The analysis does not, however, exclude the possibility that part of the functional unit may involve sites upon the myosin filament, these sites being in some way induced *in vivo*.

It is also important to point out that the value of recorded isometric tension, as is suggested by these calculations, is not seriously affected by compliance. Estimates of the total compliance for these single fibres, based on isotonic quick-release experiments from full isometric tetanus (Hoyle & Abbott, 1967), suggest that for tensions up to approx. $0.3 P_0$, which is in the range investigated with aequorin internally, the compliance is about 5 nm (50 Å) per half sarcomere. The resting sarcomere length in barnacle is 6–7 μm. Thus calculations taking into account the compliance together with the relatively slow rate of development of tension suggest that 'true' tension in barnacle over this range (active state?) is translated in front of the recorded tension by a matter of a few ms. The translation is therefore of the same order as was observed between the free calcium change and the time-course of the aequorin light emission (Ashley et al., 1973b).

In addition to predicting the tension response in the transient state, the model makes accurate predictions about tension (ATPase) and calcium bound in the steady state. In general, tension (and ATPase) rises to a maximum over about 1 pCa unit (pCa = $-\log_{10} [Ca^{2+}]$), whereas the calcium bound increases to a maximum over approximately 2 pCa units (Weber et al., 1964; Hellam & Podolsky, 1969; Julian, 1971; Kendrick-Jones et al., 1970). There was not, apparently, any simple relationship between the two phenomena, although implicitly both must be related. However, both the independent and consecutive schemes, involving the co-operative interaction of two calcium ions per functional unit, can provide a ready explanation for this apparent discrepancy. Thus from eqn. (2) for the independent model, tension and ATPase are dependent upon the product of the concentration of the sites occupied with calcium, whereas calcium bound goes simply as the sum of concentration of these sites (eqn. 3). This has been investigated in relation to experimentally determined values of myofibrillar ATPase and calcium bound measured under identical conditions (Weber et al., 1964). The results are indicated in Fig. 11 and the co-operative action of two calcium ions gives a reasonable interpretation of the experimentally determined values (for additional details see the legend to Fig. 11). This calculation is, of course, independent of the value of the total calcium bound to the myofibrils. The amount of calcium bound on the myofibrils varies even for vertebrates, for rabbit from values of about 2 μmol/g of protein (Weber et al., 1964; Weber

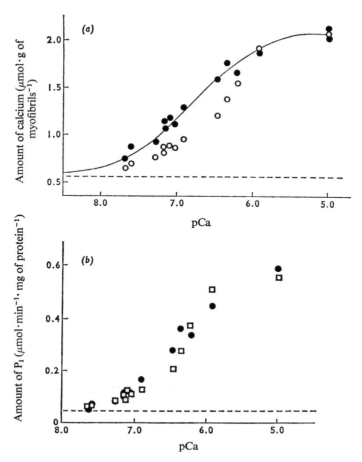

Fig. 11. (a) *Calcium bound to rabbit myofibrils as a function of pCa.* (b) *ATPase activity of rabbit myofibrils measured under the same conditions as (a), as a function of pCa*

(a) Experimental points (●) from Weber *et al.* (1964); ——, the relationship predicted by the independent or consecutive scheme at the point of equivalence; ○, calcium bound in a simultaneous scheme (see Ashley & Moisescu, 1972*b*); ----, residual calcium bound at pCa >9. (b) Experimental points (□) from Weber *et al.* (1964); ●, calcium bound experimentally, transposed from (a) into ATPase activity by using eqn. (2) and the relationship $CaM_i = c_i K_i Ca/(1 + K_i Ca)$ for the independent scheme, where $K_i = K_1 = K_2 = 8 \times 10^6 \, M^{-1}$ are the equilibrium constants. ----, Residual ATPase activity at pCa >9 (from Ashley & Moisescu, 1972*b*).

& Bremel, 1970; Bremel & Weber, 1972) to roughly half this value for frog and chicken and other species with actin-linked regulatory systems such as crustaceans (Lehman *et al.*, 1972).

Finally, experiments with single amphibian muscle fibres injected with aequorin apparently confirm the time-relations observed between free calcium and tension observed in crustacean fibres, as well as the calcium time-course deduced from the use of murexide (Fig. 6) (Rüdel & Taylor, 1973).

Mechanism of the Free Calcium Change in the Sarcoplasm

The analysis of the relationship between free calcium concentration and the development of tension provides implicitly the rate of variation of the calcium bound to the myofibrils $(dCa/dt)_{myofibril}$ and hence the myofibrillar contribution to the rate of variation of the free sarcoplasmic calcium, providing that a value of c_i is known (Ashley & Moisescu, 1973a). The time-course of these components is illustrated in Figs. 12(a) and 12(b) for the same long-duration-stimulus record as Fig. 10. In Fig. 12 trace 1 is the time-course of the free calcium determined from the aequorin light emission (from Fig. 10, trace 1), and trace 2 is essentially

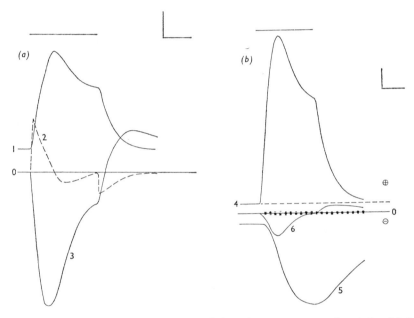

Fig. 12. (a) and (b) time-relations during a single long-duration response of an isolated Balanus muscle fibre of: trace 1, the free calcium derived from the aequorin light emission (Fig. 10, trace 1b); trace 2, $(1+A)(dCa/dt)_{free}$ where A is a constant of the order of unity and accommodates the rapid low-affinity calcium-binding sites $(1 \times 10^3 - 1 \times 10^4 \mathrm{M}^{-1})$; trace 3, $(dCa/dt)_{myofibrils\,(M)}$; trace 4, $(dCa/dt)_{sarcoplasmic\;reticulum\;release\;(SRr)} + (dCa/dt)_{leakage\;(L)}$; trace 5, $(dCa/dt)_{sarcoplasmic\;reticulum\;uptake\;(SRu)}$; trace 6, $(dCa/dt)_{sarcoplasmic\;reticulum\;sites\;(SRS)}$

●, Values of the expression F, where $F = -(1+A)(dCa/dt)_{free} + (dCa/dt)_M + (dCa/dt)_{SRr} + (dCa/dt)_{SRu} + (dCa/dt)_{SRS} + (dCa/dt)_L$. The bars represent 1% error of the expressions calculated by numerical integration (traces 3, 4, 5 and 6). +, Release of calcium into the sarcoplasm; −, uptake. Parameters used: $(1+A)$ $Ca_r = 10^{-7}\mathrm{M}$; $2c_{1,2} = 10^{-4}\mathrm{M}$; $aCa_r = 2.0368 \times 10^{-5}\mathrm{M\cdot s^{-1}}$; $b = -1.874 \times 10^{-5}\mathrm{M\cdot s^{-1}}$; $D = 8.544 \mathrm{s^{-1}}$ since $(dCa/dt)_{SRr} = B\exp(-Dt)$ when stimulation ceases, and B is the value of the regenerative release at this point in time; $(Ca_r)^2\,\alpha/\beta = 0.03$; $\beta = 2.15 \mathrm{s^{-1}}$; $E = 1.68 \times 10^{-4}\mathrm{M\cdot s^{-1}}$, $c_s = 5 \times 10^{-6}\mathrm{M}$. Calibration bar, vertical: (trace 1, Ca_r; trace 2, $2 \times 10^{-6}\mathrm{M\cdot s^{-1}}$; trace 3–trace 6 and for points (●), $1 \times 10^{-5}\mathrm{M\cdot s^{-1}}$; horizontal: 200 ms. Temperature 10–12°C. Solid horizontal lines represent the duration of the stimulus (from Ashley & Moisescu, 1973a). a and b depend upon the intensity of the depolarizing pulse (see the text); E is the maximum rate of absorption by the sarcoplasmic reticulum and is achieved when (2CaS) reaches its maximum value, c_s (see the text); D is a constant; α and β are the forward and reverse rate constants for the formation of (2CaS) and $2c_{1,2}$ is total concentration of the myofibrillar calcium-binding sites (see the text).

$(dCa/dt)_{free}$, the rate of variation of the free calcium in the fibre sarcoplasm. Tension was calculated from the model (eqns. 1 and 2) as before for the best fit to the recorded trace (Fig. 10, trace 4) and this provided not only the time-course of calcium bound, but also the component $(dCa/dt)_{myofibril}$ (Fig. 12a, trace 3). The first point that arises from the time-course of this component is that the myofibrils, during the early stages of tension development, make a major contribution to the rate of removal of free calcium from the sarcoplasm. Subsequent analysis indicates that this component makes at least an equal contribution to the removal of free calcium, as does the sarcoplasmic reticulum at this stage (Ashley & Moisescu, 1973a). This implies that in isolated myofibrillar preparations, the relatively slow rise of tension observed when the bundles are activated by externally applied calcium buffer solutions (Hellam & Podolsky, 1969; Julian, 1971) could simply be due to a limitation of the intrafibrillar free calcium. If more adequate calcium buffering conditions are provided, the tension responses of bundles of myofibrils, isolated from both amphibian and crustacean fibres, can be made to rise considerably faster than had been previously reported. The rate of rise of bundles of barnacle myofibrils approaches that observed *in vivo* (Ashley & Moisescu, 1973b).

The second point that arises from the time-course of the myofibrillar component (Fig. 12a, trace 3) is that, at a time when the free calcium in the sarcoplasm is close to its resting value [and $(dCa/dt)_{free}$ is close to zero], the rate of release of calcium from the myofibrils is still high, close to its maximum value. Thus the rate of absorption of calcium by the sarcoplasmic reticulum, $(dCa/dt)_{sarcoplasmic\ reticulum\ uptake}$, must be of the same order as the sum of the myofibrillar and leakage components $(dCa/dt)_{leakage}$ at this point. Since steady-state experiments with isolated sarcoplasmic reticulum vesicles *in vitro* indicate that calcium controls the rate of calcium uptake by the sarcoplasmic reticulum, then there must be, in the transient state, a lag between the time-course of the free calcium change and the time-course of the rate of sarcoplasmic reticulum calcium uptake. This implies that a slowly equibrating calcium-dependent step must be involved in this process. A scheme for the rate of calcium uptake by the sarcoplasmic reticulum that is consistent with the experimental results is:

$$2Ca^{2+} + S \underset{\beta}{\overset{\alpha}{\rightleftharpoons}} (2CaS) \sim \text{Sarcoplasmic reticulum ATPase}$$

$$\sim (dCa/dt)_{sarcoplasmic\ reticulum\ uptake}$$

where S is the concentration of free and c_s is the total concentration of the high-affinity sarcoplasmic reticulum calcium-binding ATPase sites and $c_s = S + (2CaS)$, so that $(dCa/dt)_{sarcoplasmic\ reticulum\ uptake} = E(2CaS)/c_s$. Other details are in the legend to Fig. 12 (see also Ashley & Moisescu, 1973a; Ashley et al., 1973b).

The main conclusions from this analysis are that the process of calcium release from the sarcoplasmic reticulum should be a regenerative one (see Ford & Podolsky, 1970, 1972; Endo et al., 1970) during stimulation [Fig. 12b, trace 4 illustrates the time-course of $(dCa/dt)_{sarcoplasmic\ reticulum\ release}$ where $(dCa/dt)_{sarcoplasmic\ reticulum\ release} = aCa_{free} + b$]; whereas after stimulation ceases, the release decays quasi-exponentially. Secondly, as mentioned above, the rate of calcium uptake by the sarcoplasmic reticulum should vary slowly with changes in free calcium concentration and the time-course of this component is shown in Fig. 12(b), trace 5.

Much of this work has been supported by grants from the Medical Research Council. We thank Dr. A. G. Lowe and Dr. D. G. Moisescu for useful discussions, and the Editors of the *Journal of Physiology, Nature* and *Biochemical and Biophysical Research Communications* for permission to reproduce published work.

References

Ashley, C. C. (1967) *Amer. Zool.* **7**, 647–659
Ashley, C. C. (1970) *J. Physiol. (London)* **210**, 133P–134P
Ashley, C. C. (1971) *Endeavour* **30**, 18–25
Ashley, C. C. & Ellory, J. C. (1972a) cited by Ashley, C. C., Caldwell, P. C. & Lowe, A. G. (1972) *J. Physiol. (London)* **223**, 735–755
Ashley, C. C. & Ellory, J. C. (1972b) *J. Physiol. (London)* **226**, 653–674
Ashley, C. C. & Ellory, J. C. (1973) *Biochem. Soc. Trans.* **1**, 96
Ashley, C. C. & Moisescu, D. G. (1972a) *J. Physiol. (London)* **226**, 82P–84P
Ashley, C. C. & Moisescu, D. G. (1972b) *Nature (London) New Biol.* **237**, 208–211
Ashley, C. C. & Moisescu, D. G. (1973a) *J. Physiol. (London)* **231**, 23P–25P
Ashley, C. C. & Moisescu, D. G. (1973b) *J. Physiol. (London)* **233**, 8P–9P
Ashley, C. C. & Ridgway, E. B. (1968) *Nature (London)* **219**, 1168–1169
Ashley, C. C. & Ridgway, E. B. (1970) *J. Physiol. (London)* **209**, 105–130
Ashley, C. C., Caldwell, P. C. & Lowe, A. G. (1972) *J. Physiol. (London)* **223**, 735–755
Ashley, C. C., Ellory, J. C. & Hainaut, K. (1973a) *J. Physiol. (London)* nthe press
Ashley, C. C., Moisescu, D. G. & Rose, R. M. (1973b) in *Calcium-binding Proteins* (Drabikowski, W., ed.), in the press, Elsevier, Amsterdam
Baker, P. F. (1972) *Progr. Biophys. Mol. Biol.* **24**, 177–221
Baker, P. F., Hodgkin, A. L. & Ridgway, A. L. (1971) *J. Physiol. (London)* **218**, 709–755
Baskin, R. J. (1971) *J. Cell Biol.* **48**, 49–60
Blaustein, M. P. & Hodgkin, A. L. (1969) *J. Physiol. (London)* **200**, 497–527
Bremel, R. D. & Weber, A. (1972) *Nature (London) New Biol.* **238**, 97–101
Caldwell, P. C. (1958) *J. Physiol. (London)* **142**, 22–62
Caldwell, P. C. (1970) in *Calcium and Cellular Function* (Cuthbert, A., ed.), pp. 10–16, Macmillan, London
Caldwell, P. C. & Walster, G. E. (1963) *J. Physiol. (London)* **169**, 353–372
Drabikowski, W., Nowak, E., Barylko, B. & Drabowska, R. (1973) *Cold Spring Harbor Symp. Quant. Biol.* **37**, 245–249
Ebashi, S., Endo, M. & Ohtsuki, I. (1969) *Quart. Rev. Biophys.* **2**, 351–384
Endo, M., Tanaka, M. & Ogawa, Y. (1970) *Nature (London)* **228**, 34–36
Fatt, P. & Katz, B. (1953) *J. Physiol. (London)* **120**, 171–204
Ford, L. E. & Podolsky, R. J. (1970) *Science* **167**, 58–60
Ford, L. E. & Podolsky, R. J. (1972) *J. Physiol. (London)* **223**, 1–19
Greaser, M. L., Yamaguchi, M., Brekke, C., Potter, J. & Gergely, J. (1973) *Cold Spring Harbor Symp. Quant. Biol.* **37**, 235–244
Harigaya, S. & Schwartz, A. (1969) *Circ. Res.* **25**, 781–794
Harigaya, S., Ogawa, Y. & Sugita, H. (1968) *J. Biochem. (Tokyo)* **63**, 324–331
Hastings, J. W., Mitchell, G., Mattingly, P. H., Blinks, J. R. & van Leeuwen, N. (1969) *Nature (London)* **222**, 1047–1050
Heilbrunn, L. V. & Wiercinski, F. J. (1947) *J. Cell. Comp. Physiol.* **29**, 15–32
Hellam, D. C. & Podolsky, R. J. (1969) *J. Physiol. (London)* **200**, 807–819
Hill, A. V. (1948) *Proc. Roy. Soc. Ser. B* **135**, 446–453
Hill, A. V. (1949) *Proc. Roy. Soc. Ser. B* **136**, 399–420
Hoyle, G. & Abbott, B. C. (1967) *Amer. Zool.* **7**, 611–614
Hoyle, G. & Smyth, T. (1963) *Comp. Biochem. Physiol.* **10**, 291–314
Hoyle, G., McNeil, P. A. & Selverston, A. I. (1972) *J. Cell Biol.* **56**, 74–91
Inesi, G. & Scarpa, A. (1972) *Biochemistry* **11**, 356–359
Izutsu, K. T., Felton, S. P., Siegel, I. A., Yoda, W. T. & Chen, A. C. N. (1972) *Biochem. Biophys. Res. Commun.* **49**, 1034–1039
Jöbsis, F. F. & O'Connor, M. J. (1966) *Biochem. Biophys. Res. Commun.* **25**, 246–252

Julian, F. J. (1971) *J. Physiol. (London)* **218**, 117–146
Kendrick-Jones, J., Lehman, W. & Szent-Györgyi, A. G. (1970) *J. Mol. Biol.* **54**, 313–326
Lehman, W., Kendrick-Jones, J. & Szent-György, A. G. (1973) *Cold Spring Harbor Symp. Quant. Biol.* **37**, 319–330
Loschen, G. & Chance, B. (1971) *Nature (London) New Biol.* **233**, 273–274
Ohnishi, T. & Ebashi, S. (1963) *J. Biochem. (Tokyo)* **54**, 506–511
Ohnishi, T. & Ebashi, S. (1964) *J. Biochem. (Tokyo)* **55**, 599–603
Perry, S. V., Cole, H. A., Head, J. F. & Wilson, F. J. (1973) *Cold Spring Harbor Symp. Quant. Biol.* **37**, 251–262
Portzehl, H., Caldwell, P. C. & Rüegg, J. C. (1964) *Biochim. Biophys. Acta* **79**, 581–591
Ridgway, E. B. & Ashley, C. C. (1967) *Biochem. Biophys. Res. Commun.* **29**, 229–234
Ringer, S. (1883) *J. Physiol. (London)* **4**, 29–42
Rüdel, R. & Taylor, S. R. (1973) *J. Physiol. (London)* **233**, 5P–6P
Sandow, A. (1952) *Yale J. Biol. Med.* **25**, 176–201
Selverston, A. I. (1967) *Amer. Zool.* **7**, 515–525
Shimomura, O. & Johnson, F. H. (1969) *Biochemistry* **8**, 3991–3997
Shimomura, O. & Johnson, F. H. (1970) *Nature (London)* **227**, 1356–1357
Shimomura, O., Johnson, F. H. & Saiga, Y. (1962) *J. Cell. Comp. Physiol.* **59**, 223–239
Shimomura, O., Johnson, F. H. & Saiga, Y. (1963) *J. Cell. Comp. Physiol.* **62**, 1–8
van Leeuwen, M. & Blinks, J. R. (1969) *Fed. Proc. Fed. Am. Soc. Exp. Biol.* **28**, Abstr. 571
Weber, A. & Bremel, R. D. (1970) in *Contractility of Muscle Cells and Related Processes* (Podolosky, R. J., ed.), pp. 37–45, Prentice Hall, London
Weber, A., Herz, R. & Reiss, I. (1964) *Proc. Roy. Soc. Ser. B* **160**, 489–501

The Role of Phosphorylase Kinase in the Nervous and Hormonal Control of Glycogenolysis in Muscle

By PHILIP COHEN

Department of Biochemistry, Medical Sciences Institute, University of Dundee, Dundee DD1 4HN, U.K.

Synopsis

The total dependence of phosphorylase kinase activity on calcium ions synchronizes glycogenolysis with the onset of muscle contraction. The activity is also stimulated by phosphorylation of the enzyme, a reaction catalysed by cyclic AMP-dependent protein kinase and also by phosphorylase kinase itself. Phosphorylase kinase possesses the structure $(\alpha\beta\gamma)_4$, where the molecular weights of the three protein subunits α, β and γ are 145000, 130000 and 45000 respectively. The 40-fold activation produced by cyclic AMP-dependent protein kinase is paralleled by the phosphorylation at a single site on the β-subunit, although one site on the α-subunit is also phosphorylated at a 5-fold slower rate, without affecting the total activity. The phosphorylation of the α-subunit does, however, greatly stimulate the dephosphorylation of the β-subunit and inactivation of the enzyme by phosphorylase kinase phosphatase, suggesting a role in the regulation of the dephosphorylation reaction. Phosphorylase kinase may also be activated up to 100-fold by limited proteolysis, but in contrast with activation by phosphorylation this appears to correlate with a modification of the α-subunit. Models for the interrelationship between the various activating mechanisms are considered, and the roles of the three subunits are discussed in terms of electron micrographs of the enzyme molecule.

Introduction

Skeletal muscle contraction relies predominantly on the mobilization of glycogen for energy once the creatine phosphate stores have been exhausted. Glycogen degradation is catalysed by two enzymes, phosphorylase and debranching enzyme (Cori & Larner, 1951). The former catalyses the sequential phosphorolysis of α1–4-linked glucose units to glucose 1-phosphate until four glucose units remain before an α1–6 branch point, whereas debranching enzyme containing both oligo-1,4–1,4-glucantransferase activity (Brown & Illingworth, 1962) and α1–6-glucosidase activity, then eliminates the 1–6 branch point, allowing the action of phosphorylase to continue.

The rate of glycogen degradation must therefore be at least partially controlled by phosphorylase because the debranching enzyme cannot act in its absence, and it is now well established that in resting muscle very little breakdown of glycogen

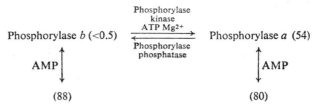

Scheme 1. *Activities of phosphorylases b and a in the presence and absence of AMP*

The numbers represent enzyme activities (units/mg) of the rabbit skeletal muscle enzyme assayed in the direction of glycogen synthesis in 0.1 M-maleate buffer, pH 6.5, at 1% glycogen, 75 mM-glucose 1-phosphate and 1.0 mM-AMP, where added (taken from Cohen *et al.*, 1971).

occurs because phosphorylase is present in an inactive '*b*' form. Phosphorylase *b* may be activated in two different ways:

(1) by AMP (Cori & Cori, 1936)

(2) by conversion into active phosphorylase *a*, a reaction catalysed by phosphorylase kinase in the presence of ATP and magnesium ions (Krebs & Fischer, 1956).

These two mechanisms each enhance phosphorylase *b* activity more than 100-fold *in vitro*, and are considered to bring about similar changes in conformation, as the two effects are not additive (Scheme 1).

Activation of Phosphorylase Kinase by Calcium Ions

Phosphorylase is totally in the inactive *b* form in resting muscle, suggesting that phosphorylase kinase activity is also switched off under these conditions. Electrical excitation of frog sartorious muscle brings about conversion of phosphorylase *b* into *a* with a half-time of about 1 s (Danforth *et al.*, 1962), suggesting that there is a very rapid method of activating this enzyme after the onset of muscle contraction. Soon after the discovery of phosphorylase kinase (Krebs & Fischer, 1956) the possibility that this activation might be exerted through calcium ions was considered, but the situation remained confused for some time because of the complication of a second effect of calcium ions on the activity of the enzyme.

Irreversible Activation by Calcium Ions

Phosphorylase kinase was initially shown to be activated 50–100-fold at physiological pH by calcium ions in the presence of a further protein factor (Meyer *et al.*, 1964). This protein was purified 3000-fold from rabbit muscle extracts and shown to be a proteolytic enzyme dependent on calcium ions for activity (Huston & Krebs, 1968). The irreversible nature of this process and the requirement for relatively high calcium concentrations suggested this effect was not responsible for the very rapid phosphorylase *b*-into-*a* conversion that occurs after electrical excitation of muscle. It is also presumed that this proteinase is lysosomal in origin, which would preclude its action under normal cellular conditions, but this has not yet been demonstrated. The great susceptibility of phosphorylase kinase to proteolytic activation also leads to some reservations concerning the exact mechanisms of control of this enzyme *in vivo*, as discussed below.

Reversible Activation by Calcium Ions

The introduction of the relatively specific calcium chelator EGTA [ethane-dioxybis(ethylamine)tetra-acetic acid], led to the discovery of a second reversible effect of calcium ions operating at much lower concentration. Phosphorylase kinase activity was found to be considerably decreased when either EGTA was added or calcium-free solutions were used, and this inhibition was reversed by the addition of calcium ions (Ozawa et al., 1967; Brostrom et al., 1971a). Half-maximal activation occurred at approximately 1 μM-calcium ion concentration. In an elegant experiment, it was shown that the phosphorylase kinase catalysed conversion of phosphorylase b into a in vitro could be arrested instantaneously by the addition of elements of the sarcoplasmic reticulum, and reversed by addition of an excess of calcium ions (Brostrom et al., 1971a). Thus the naturally occurring calcium chelator in muscle can compete successfully for the calcium, which is tightly bound to phosphorylase kinase, suggesting the process is rapidly reversible in vivo.

A complete dependence of phosphorylase kinase activity on calcium ions was also demonstrated by using glycogen particles isolated from rabbit muscle in the presence of EDTA, which contain substantial quantities of the phosphorylase, phosphorylase kinase, phosphorylase phosphatase and glycogen synthetase of the cell (Meyer et al., 1970; Heilmeyer et al., 1970). Addition of ATP and magnesium ions failed to convert any phosphorylase b into a unless calcium ions were added, and the concentration of calcium ions required for half maximal activation was again 1 μM. The results therefore suggested that calcium ions provide the link between the onset of muscle contraction (Ebashi et al., 1969) and initiation of glycogenolysis by phosphorylase kinase.

Isolation of Phosphorylase Kinase

Phosphorylase kinase occupies an important position in muscle cell metabolism. It is the point at which the nervous control of glycogenolysis is exerted through calcium ions, and as will be discussed below it is also subject to hormonal control through a cyclic AMP-dependent phosphorylation and activation of the protein. Therefore the interrelationship of calcium ions and phosphorylation in this protein represents the balance between the effects of two critical physiological stimuli. Phosphorylase kinase is also unique in participating directly in the two kinase–phosphatase couples that contribute to the regulation of glycogenolysis.

Phosphorylase kinase represents approximately 0.8% of the soluble protein in rabbit muscle (Delange et al., 1968; Cohen, 1973) and its concentration in rabbit muscle would be close to 1 mg/ml assuming it is uniformly distributed within the muscle cell. The very high concentration for this activating enzyme, almost 15% by weight of its substrate phosphorylase, was initially surprising, but conventional criteria of protein purity indicate that preparations exhibit a high degree of homogeneity. Such tests have included sedimentation velocity and sedimentiaton equilibrium experiments (Hayakawa et al., 1973a; Cohen, 1973), chromatography on DEAE-cellulose, gel filtration and antigen–antibody precipitation in agar (Cohen, 1973) and polyacrylamide-gel electrophoresis (Hayakawa et al., 1973a).

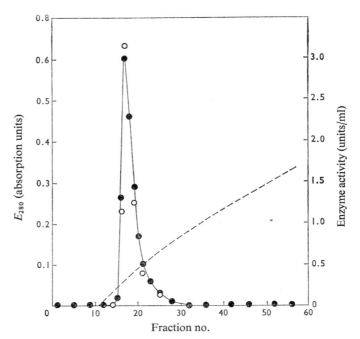

Fig. 1. *Chromatography of purified phosphorylase kinase on a DEAE-cellulose column (20 cm × 1 cm) equilibrated with 50 mM-sodium glycerophosphate–2 mM-EDTA–1 mM-dithiothreitol, pH 7.0*
●, E_{280}; ○, enzyme activity; ----, NaCl concentration (M). Enzyme (15 mg) was applied to the column and the recovery of protein was 90% (reproduced from Cohen, 1973).

Fig. 1 shows DEAE-cellulose chromatography of the purified enzyme, which yields a single protein peak coinciding with enzyme activity. This is a particularly good indication of purity, as no separation based on charge is used in the isolation procedure (Cohen, 1973).

Subunit Structure of Phosphorylase Kinase

Although phosphorylase kinase appears homogeneous by a number of tests, gel electrophoresis under dissociating conditions reveals four bands staining for protein termed α, α′, β and γ (Plate 1) (Cohen, 1972, 1973). Phosphorylase kinase has a strong tendency to polymerize to active molecular aggregates, which may be separated by gel filtration. These aggregated forms show all four components, and all four are also retained after chromatography of the purified enzyme on DEAE-cellulose (Fig. 1). Carboxymethylation with iodoacetate, performic acid oxidation and dissolution in boiling 6 M-guanidinium chloride followed by dialysis against sodium dodecyl sulphate also yield an identical pattern, suggesting that the result is not caused by incomplete reduction and denaturation of the enzyme. Addition of a variety of inhibitors of proteolytic enzymes at each step of the purification procedure and before sodium dodecyl sulphate denaturation also have no influence on the gel pattern, suggesting that the four components are not derived from one another by limited proteolysis. These inhibitors include di-isopropyl-fluorophosphate and phenylmethylsulphonyl fluoride to inhibit serine proteinases, tosyl-lysylchloromethylketone to inhibit thiol proteinases such as cathepsin B1, and pepstatin to inhibit cathepsins D and E (Barrett & Dingle, 1972). Gentle stirring of minced rabbit muscle in iso-osmotic sucrose to decrease lysosomal rupture rather than the normal homogenization in dilute EDTA solutions, is also ineffectual in altering the gel pattern obtained with the purified enzyme. Nevertheless, not all the lysosomal proteinases would be inhibited by the reagents used, and the presence of the faint component termed α' suggests that a slight proteolysis of the α component might have occurred. This is further supported by the isolation of phosphorylase kinase from normal murine muscle in which the relative intensity of α and α' is reversed (P. T. Cohen, unpublished work).

The molecular weights of the four components estimated empirically by gel electrophoresis are $\alpha = 145000$, $\alpha' = 140000$, $\beta = 128000$ and $\gamma = 41000$. Separation of the γ subunit from the $\alpha + \alpha' + \beta$ subunits (Plate 2) followed by sedimentation equilibrium in the presence of 6M-guanidinium chloride gave $\alpha + \alpha' + \beta = 137000 \pm 5000$ and $\gamma = 47000 \pm 2000$. The average of these two methods of molecular-weight determination suggested the values $\alpha = 145000$, $\beta = 130000$ and $\gamma = 45000$. The amino acid composition of the γ subunit shows significant differences from the α and β subunits (Cohen, 1973).

The molar ratios of the three subunits α, β and γ were determined from the above molecular weights, and weight ratios of the subunits estimated in three different ways; (a) from densitometric tracings of the polyacrylamide gels (Plate 1), (b) by gel filtration on Sephadex G-200 in the presence of sodium dodecyl sulphate (Plate 2a) and (c) by the incorporation of iodo[^{14}C]acetate into each subunit after complete carboxymethylation and using the known thiol content of the subunits (Cohen, 1973). Each method suggested that the α, β and γ subunits were present in equimolar quantities, giving a minimum binding weight, $\alpha\beta\gamma$, of 320000.

Hayakawa et al. (1973a,b) have reported lower molecular weights of 118000 and 108000 for the α and β chains and a molar ratio $\alpha:\beta:\gamma$ of 1:1:2. These molecular weights were obtained by dodecyl sulphate gel electrophoresis and supported by sedimentation equilibrium at pH 10.3 and in the presence of dodecyl sulphate at pH 9. In the former procedure the dimer of serum albumin was used as the standard marker in the molecular-weight range of the α and β subunits, while in the sedimentation equilibrium experiments the molecular weights may possibly be underestimated because of the problems of charge effects far from the isoelectric point, and uncertain decreases in apparent specific volume due to dodecyl sulphate binding. In the present work, the higher molecular weights obtained by dodecyl sulphate gel electrophoresis for the α and components relied on the β and β' subunits of RNA polymerase, β-galactosidase and glycogen phosphorylase as internal standards, and these values were supported by sedimentation equilibrium in the presence of 6M-guanidinium chloride (Cohen, 1973). The $\alpha:\beta:\gamma$ ratio of 1:1:2 obtained by densitometric tracings of the polyacrylamide gels (Hayakawa et al., 1973a) is also an unresolved discrepancy with the results given in this article in which the molar proportions $\alpha:\beta:\gamma$ were 1:1:1 by three methods (Cohen, 1973).

The molecular weight of the native enzyme determined by sedimentation equilibrium was 1280000±20000, demonstrating that the smallest active species of phosphorylase kinase contains 4.0 $\alpha\beta\gamma$ units.

It has been reported that phosphorylase kinase binds 1 mol of calcium/ 57000g of protein (Brostrom et al., 1971a), which corresponds to approximately six calcium molecules/$\alpha\beta\gamma$ moiety, or 24/enzyme molecule. The concentration of phosphorylase kinase in rabbit skeletal muscle is 3μM in terms of $\alpha\beta\gamma$ units; therefore 18μM-calcium would be bound to phosphorylase kinase if it were fully saturated.

Hormonal Control of Phosphorylase Kinase

The finding that adrenalin promoted the formation of phosphorylase a in liver (Rall et al., 1956) was the original observation in this system that led to the discovery of adenylate cyclase and cyclic AMP (Rall et al., 1957). Highly purified phosphorylase kinase was found to become phosphorylated and activated some twentyfold by preincubation with cyclic AMP, ATP and magnesium (DeLange et al., 1968), but no cyclic AMP binding to the enzyme occurred under these conditions. This led to the discovery of a further enzyme present as a trace contaminant in phosphorylase kinase preparations, which has been termed 'cyclic AMP-dependent protein kinase' (Walsh et al., 1968, 1971a). The activation and phosphorylation of phosphorylase kinase is however catalysed by two different mechanisms:

(1) by cyclic AMP-dependent protein kinase in the presence of cyclic AMP, ATP and magnesium ions.

(2) by phosphorylase kinase itself at high magnesium and ATP concentrations in the absence of cyclic AMP.

These two effects were clearly distinguished from one another by making use of a heat- and acid-stable protein in rabbit skeletal muscle, which is a specific inhibitor of cyclic AMP-dependent protein kinase (Walsh et al., 1971b; Ashby & Walsh, 1972) to inhibit reaction (1) and EGTA to block reaction (2). The detailed analysis of these two activation reactions is considered below.

Cyclic AMP-Dependent Protein Kinase

This enzyme has been obtained in a homogeneous form from skeletal muscle (Brostrom et al., 1971b) and heart muscle (Rubin et al., 1972). It is composed of two types of polypeptide chain, a regulatory subunit that binds cyclic AMP, and a catalytic subunit that carries the active site. The two subunits dissociate upon addition of cyclic AMP, and the free catalytic subunit is termed 'cyclic AMP-independent protein kinase' (Krebs, 1972). The concentration of cyclic AMP required for half-maximal activation is below 0.1μM.

The cyclic AMP-dependent protein kinase from rabbit skeletal muscle was separated into two major fractions by chromatography on DEAE-cellulose, termed 'peak I' and 'peak II' (Reimann et al., 1971). Peak I has recently been further subdivided into three active species, which differ in the size of their regulatory subunits (Corbin et al., 1972), and it has been suggested that these 'isoenzymes' might be derived from one another by a limited proteolysis of the

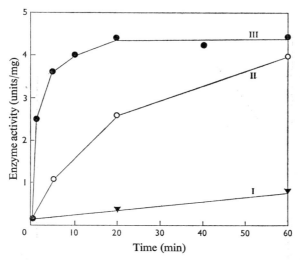

Fig. 2. *Activation of phosphorylase kinase at pH 6.8 and 25°C*
Curve (I): phosphorylase kinase + Mg–ATP; curve (II): phosphorylase kinase + Mg–ATP + cyclic AMP; curve (III): phosphorylase kinase + Mg–ATP + cyclic AMP + cyclic AMP-dependent protein kinase (reproduced from Cohen, 1973).

regulatory subunit. Both peaks I and II behave identically in the phosphorylation of the subunits of phosphorylase kinase described below (Cohen, 1973).

Phosphorylation and Activation of Phosphorylase Kinase by Cyclic AMP-Dependent Protein Kinase

Phosphorylase kinase isolated in this laboratory is characterized by a very low activity at physiological pH (6.8) compared with the activity at the pH optimum (8.2). The pH 6.8/pH 8.2 activity ratio of the freshly isolated enzyme is 0.01–0.02 (Cohen, 1973) and for the discussion at present this will be assumed to represent the totally dephosphorylated enzyme. Maximal activation obtained upon addition of cyclic AMP, ATP, Mg^{2+} and cyclic AMP-dependent protein kinase is 25–50-fold at pH 6.8 and 1.4-fold at pH 8.2, yielding a pH 6.8/pH 8.2 activity ratio of 0.36 for the phosphorylated form of the enzyme (Cohen, 1973).

The behaviour observed with phosphorylase kinase upon addition of various components of the activation reaction is shown in Fig. 2. The experiments are carried out at low magnesium and ATP concentrations (0.2 mM-ATP–2.0 mM-Mg^{2+}) and in the presence of EGTA (0.2 mM) to prevent phosphorylase kinase from catalysing its own activation (Walsh *et al.*, 1971a). Addition of ATP–Mg alone results in only a very slight increase in phosphorylase kinase activity over 1 h (curve I), but the further addition of cyclic AMP (0.01 mM) produces a much greater activation of the enzyme (curve II). These two effects are attributable to trace autophosphorylation and trace contamination with cyclic AMP-dependent protein kinases respectively. It may be estimated that the activation in curve II would be produced by only a 0.002% contamination by weight with cyclic AMP-dependent protein kinase, compared with an approximately 5% weight ratio in

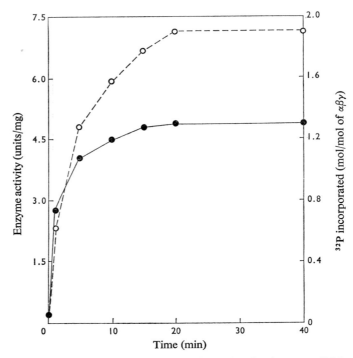

Fig. 3. *Activation and phosphorylation of phosphorylase kinase at pH 6.8*
●, Activity; ○, ^{32}P incorporated into the protein. Experimental conditions are the same as Fig. 2, curve III (reproduced from Cohen, 1973).

rabbit muscle. Further addition of small amounts of purified cyclic AMP-dependent protein kinase produces a much more rapid activation, which reaches a maximum within 20 min (curve III, Fig. 2).

Analysis of the ^{32}P incorporated into the enzyme from [γ-^{32}P]ATP shows a plateau at very close to two molecules of phosphate/$\alpha\beta\gamma$ group, although phosphorylation and activation do not exactly parallel one another (Fig. 3). The analysis of ^{32}P incorporated into each subunit shows that activation parallels the incorporation of phosphate into just one site on the β subunit, while a second site, on the α subunit is phosphorylated at a slower rate without any apparent effect on activity. Thus after 1 min, both activation and phosphorylation of the β subunit were approximately 60% complete, while the α subunit was only phosphorylated to 6% (Fig. 4). After 20 min both α and β subunits contain 1 molecule of covalently bound phosphate. The stoicheiometry of the reaction provides further evidence for the equimolar proportions of the α and β subunits, and for the $\alpha\beta\gamma$ binding weight of 320000. No radioactivity appeared in the γ subunit.

Role of α-Subunit Phosphorylation

(a) Specificity of cyclic AMP-dependent protein kinase

The results described above suggested that the activation of phosphorylase kinase by cyclic AMP-dependent protein kinase was a specific process, associated

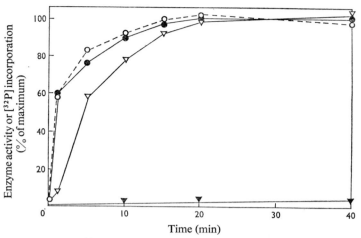

Fig. 4. *Incorporation of ^{32}P into the α subunit ($\triangledown$), the β subunit ($\bullet$) and the γ subunit ($\blacktriangledown$) upon addition of Mg–ATP, cyclic AMP and cyclic AMP-dependent protein kinase*
○, Enzyme activity (reproduced from Cohen, 1973).

with a unique site on the β subunit. However, the phosphorylation of a second site on the α subunit without any apparent increase in activity, at a slower, but still rapid rate, posed the question of whether it might play any role in the hormonal control of phosphorylase kinase.

Cyclic AMP-dependent protein kinase has also been shown to phosphorylate and inactivate glycogen synthetase (Soderling *et al.*, 1970). The same enzyme can therefore stimulate glycogen breakdown and inhibit glycogen synthesis. It is also capable of phosphorylating casein, histone, protamine and phosvitin (Walsh *et al.*, 1968; Reimann *et al.*, 1971), which has led to the suggestion that the enzyme

Table 1. *Phosphorylation of glycolytic enzymes by cyclic AMP-dependent protein kinase (peak II from DEAE-cellulose; Cohen, 1973)*
The enzymes were from rabbit skeletal muscle, except fructose 1,6-diphosphatase, which was from rabbit liver. Reactions were carried out in 10mM-glycerophosphate, 0.4mM-EDTA, 0.2mM-EGTA, 0.01 mM-cyclic AMP, 2mM-magnesium acetate, 0.2mM-ATP, pH7.0 (P. Cohen, unpublished work). The phosphorylase *b* contains traces of phosphorylase kinase activity.

Enzyme	Relative activity (%)
Phosphorylase kinase	100*
Phosphorylase *b*	0.4
Phosphoglucomutase	0.03
Phosphofructokinase	0.3
Aldolase	0.01
Triose phosphate isomerase	0.01
Triose phosphate dehydrogenase	0.01
Phosphoglycerate kinase	0
Phosphoglyceromutase	0
Enolase	0.02
Pyruvate kinase	0
Lactate dehydrogenase	0.01
Creatine kinase	0.02
Adenylate kinase	0
Fructose diphosphatase	0.2

* Rate of β-subunit phosphorylation.

may be relatively non-specific. Although cyclic AMP-dependent protein kinase has a tendency to label a number of proteins at a slow rate, the data in Table 1 show that the rate of phosphorylation of any other enzyme in the glycolytic pathway is extremely slow compared with the rate of phosphorylation and activation of phosphorylase kinase. Clearly the rate of phosphorylation of the α-subunit is more than two orders of magnitude greater than the phosphorylation of any of the proteins shown in Table 1. It should also be emphasized that the potential activity of the cyclic AMP-dependent protein kinase in rabbit skeletal muscle is very high, and the time required to achieve half-maximal phosphorylation and activation of the β subunit may be estimated to be of the order of 1 s, again assuming uniform distribution of the enzyme within the cell.

(b) *Phosphorylation and dephosphorylation of phosphorylase kinase in glycogen particles*

Perfusion of rat skeletal muscle with adrenalin leads to an increase in cyclic AMP concentration and an increase in the pH 6.8/pH 8.2 activity ratio of phosphorylase kinase (Posner *et al.*, 1965; Drummond *et al.*, 1969), but a direct demonstration that the phosphorylase kinase molecule becomes phosphorylated after administration of adrenalin has not yet been achieved (Meyer & Krebs, 1970). However, the glycogen particles mentioned earlier (Meyer *et al.*, 1970), which contain substantial quantities of the enzymes of glycogen metabolism and which are thought to resemble a functional complex in rabbit skeletal muscle, provide an alternative system for investigating this problem. The particles are prepared in the presence of EDTA to exclude calcium from the system, and therefore represent the resting muscle situation. Phosphorylase and phosphorylase kinase are both in their inactive dephosphorylated forms (Heilmeyer *et al.*, 1970), although glycogen synthetase is predominantly in the phosphorylated and inactive D form (Soderling *et al.*, 1970).

The addition of cyclic AMP, ATP and Mg^{2+} fails to produce significant conversion of phosphorylase *b* into *a* in a freshly isolated particle preparation unless calcium ions are also included (Fig. 5a). Calcium ions then produce a rapid conversion of phosphorylase *b* into *a*, followed by an almost equally rapid reconversion into phosphorylase *b*, termed 'flash activation' (Heilmeyer *et al.*, 1970). The particles are known to be contaminated by elements of the sarcoplasmic reticulum possessing powerful ATPase (adenosine triphosphatase) activity; phosphorylase *b*-into-*a* conversion therefore occurs until all the ATP has been consumed, and phosphorylase phosphatase then reconverts phosphorylase *a* back into phosphorylase *b* (Heilmeyer *et al.*, 1970). If, however, phosphorylase kinase activity is examined this exhibits a similar 'flash' activation to phosphorylase but in the absence of calcium ions (Fig. 5b), suggesting that phosphorylase kinase is being phosphorylated and dephosphorylated under these conditions. A less pronounced effect is also observed in the absence of cyclic AMP (Fig. 5b). Phosphorylation of phosphorylase kinase in the absence of calcium ions without significant phosphorylase *b*-into-*a* conversion clearly shows that the phosphorylated form of phosphorylase kinase, like the dephosphorylated form is also totally dependent on calcium ions for activity. This result was also suggested by experi-

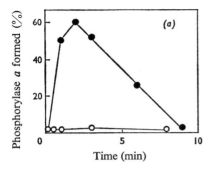 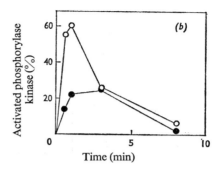

Fig. 5. *Effect of calcium ions and cyclic AMP on activation of phosphorylase and phosphorylase kinase*

(a) Formation of phosphorylase a in a glycogen-particle suspension after addition of 5mM-ATP, 20mM-magnesium and 1 μM-cyclic AMP in the presence (●) or absence (○) of 0.1 mM-Ca^{2+}. (b) Formation of activated phosphorylase kinase in the glycogen particles after addition of 5mM-ATP and 20mM-magnesium in the presence (○) or absence (●) of 0.001 mM-cyclic AMP (S. J. Yeaman & P. Cohen, unpublished work).

ments with the purified enzyme phosphorylated *in vitro*, which was 70–80% inhibited when calcium-free reagents were used during preparation and assay of the enzyme (Brostrom *et al.*, 1971a).

Polyacrylamide-gel electrophoresis of the glycogen particles in the presence of sodium dodecyl sulphate is shown in Plate 3. Relatively few protein-staining components are seen, four of which correspond to the α and β subunits of phosphorylase kinase, phosphorylase and glycogen synthetase. The phosphorylation and dephosphorylation of phosphorylase kinase during flash activation (Fig. 5b) has been substantiated by using [γ-^{32}P]ATP and gel electrophoresis to follow the labelling of the enzyme subunits. The α and β subunits are both phosphorylated to approximately the same extent during the flash activation of phosphorylase kinase (S. J. Yeaman & P. Cohen, unpublished work). Therefore in a system thought to resemble the situation *in vivo* the α subunit is again rapidly phosphorylated and dephosphorylated.

(c) Implication of α-subunit phosphorylation in the regulation of phosphorylase kinase phosphatase

Highly purified phosphorylase kinase is also contaminated with trace amounts of the enzyme that reverses the cyclic AMP-dependent activation and phosphorylation of the enzyme. The relative activities of the enzyme, termed 'phosphorylase kinase phosphatase' (Riley *et al.*, 1968) and the cyclic AMP-dependent protein kinase together control the extent of phosphorylation and activity of phosphorylase kinase. During the development of a suitable assay system for the phosphatase, a surprising observation was made, which suggested that the phosphorylation of the α subunit may play an important role in the regulation of the phosphorylase kinase dephosphorylation reaction (Cohen & Antoniw, 1973). The results of these experiments are given in Figs 6 and 7.

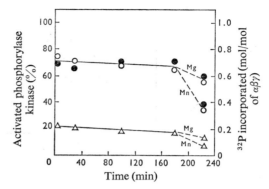

Fig. 6. *Influence of endogenous phosphorylase kinase phosphatase activity at pH 7.0 on phosphorylase kinase (2 mg/ml) containing 0.98 mol of phosphate/αβγ (α = 0.23, β = 0.75)*

The activation reaction was terminated at time $t = 0$ min by addition of an equal volume of 0.01 M-EDTA, pH 7.0, containing 80%-saturated ammonium sulphate. The solution was kept in ice for 10 min and then centrifuged at 15 000g for 2 min. The precipitate was suspended at 20°C in Tris–HCl (I 0.02)–1.0 mM-EDTA–1.0 mM-dithiothreitol, pH 7.0, containing 30% ammonium sulphate. The suspension was recentrifuged and the precipitate redissolved in the same buffer but without ammonium sulphate. The solution was next dialysed against this buffer for 1 h at 20°C and then stored for a further 1 h. Analyses for phosphorylase kinase activity (●), β-subunit phosphorylation (○) and α-subunit phosphorylation (△) were carried out at each step from $t = 0$ min. The broken lines indicate the effect of adding 10 mM-Mg^{2+} or 1 mM-Mn^{2+} at 180 min (reproduced from Cohen & Antoniw, 1973).

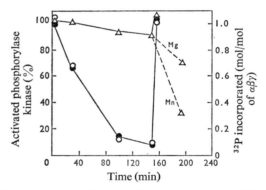

Fig. 7. *Influence of endogenous phosphorylase kinase phosphatase activity at pH 7.0 on phosphorylase kinase (2 mg/ml) containing 2.1 mol of phosphate (αβγ (α = 1.05; β = 1.05)*

Analyses for phosphorylase kinase activity (●), β-subunit phosphorylation (○) and α-subunit phosphorylation (△) were carried out. After 150 min ATP-Mg^{2+}, cyclic AMP and cyclic AMP-dependent protein kinase were added. The broken lines indicate a second experiment in which 10 mM-Mg^{2+} or 1 mM-Mn^{2+} was added at 150 min. Other experimental conditions are as in Fig. 6 (reproduced from Cohen & Antoniw, 1973).

In these experiments phosphorylase kinase was phosphorylated by cyclic AMP-dependent protein kinase until 1 phosphate (Fig. 6) or 2 phosphates (Fig. 7) had been incorporated/αβγ. As expected from the results presented earlier (Fig. 4), the β-subunit was predominantly labelled in the former reaction (Fig. 6) while the latter reaction contained close to 1 molecule of phosphate in both the α- and β-subunits/αβγ (Fig. 7). However, the rate of dephosphorylation of these two phosphoenzymes by the trace endogenous phosphorylase kinase phosphatase activity was remarkably different.

After termination of the activation reaction at 1 molecule of phosphate incorporated/$\alpha\beta\gamma$, the rate of dephosphorylation of both the α- and β-subunits and enzyme inactivation were extremely slow (Fig. 6) and required the addition of bivalent cations (180 min) to stimulate these reactions at an appreciable rate. However, when the activation reaction was terminated only after the phosphorylation of the α- and β-subunits was complete ($\alpha = 1.05$ and $\beta = 1.05$ per $\alpha\beta\gamma$), there was an immediate rapid inactivation of the enzyme and a specific dephosphorylation of the β-subunit. The perfect correlation between enzyme inactivation and loss of phosphate from the β-subunit conclusively demonstrated that the reversible activation of phosphorylase kinase catalysed by cyclic AMP-dependent protein kinase and phosphorylase kinase phosphatase depends only on the reversible phosphorylation of a unique site on the β-subunit. This therefore led to the formation of a form of the enzyme which had almost returned to the original activity before activation but still contained close to 1 covalently bound phosphate molecule in the α-subunit (Fig. 7). The readdition of cyclic AMP-dependent protein kinase, cyclic AMP, ATP and magnesium ions at this time led to a reactivation of the enzyme and a rephosphorylation of the β-subunit (Fig. 7), showing that phosphorylation of the α-subunit does not prevent the reactivation process. Dephosphorylation of the α-subunit was achieved by the addition of bivalent cations at 150 min (Fig. 7).

The 100-fold stimulation of β-subunit dephosphorylation by α-subunit phosphorylation represents a new form of enzyme regulation by covalent modification, namely the control of protein dephosphorylation by 'second-site phosphorylation' (Cohen & Antoniw, 1973). The labelling of the second site on the α-subunit presumably alters the conformation of phosphorylase kinase in such a way as to greatly enhance the reactivity of its phosphatase at the primary phosphorylated site responsible for enzyme activation. In this particular instance it is undoubtedly an effective mechanism for exerting control of the cyclic AMP-dependent protein kinase–phosphorylase kinase phosphatase enzyme cycle (Cohen & Antoniw, 1973).

Phosphorylase kinase phosphatase and glycogen synthetase phosphatase may be one and the same enzyme (Zieve & Glinsmann, 1973). Therefore both phosphorylation and dephosphorylation of phosphorylase kinase and glycogen synthetase are catalysed by the same kinase and phosphatase.

Catalysis of Phosphorylase Kinase Activation by Phosphorylase Kinase

Phosphorylase kinase is able to phosphorylate and activate itself at high ATP and magnesium ion concentration. This reaction is totally dependent on calcium ions and may be completely inhibited by EGTA (Walsh et al., 1971a). Fig. 8 shows a progress curve for the formation of phosphorylase a catalysed by the dephosphorylated form of phosphorylase kinase at pH 6.8. Phosphorylase kinase activity increases 2–3-fold during the first 5 min of the reaction, corresponding to an increase in pH 6.8/pH 8.2 activity ratio of 0.015 to 0.04. This represents approximately 8 % conversion into the phosphorylated, activated form

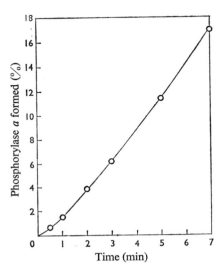

Fig. 8. *Catalysis of phosphorylase a formation by dephosphorylated phosphorylase kinase at pH 6.8, 30°C*

The assay contained phosphorylase b, 5 mg/ml, 3 mM-ATP–10 mM-magnesium acetate, and phosphorylase kinase, 0.015 mg/ml (P. Cohen, unpublished work).

of the enzyme. Over the same 5 min period, 12% of the phosphorylase b is converted into phosphorylase a. As phosphorylase kinase activity increases essentially linearly with increasing phosphorylase and phosphorylase kinase concentration, it follows that the percentage conversion of phosphorylase kinase into its activated form should be comparable with the percentage conversion of phosphorylase b into phosphorylase a after activation by calcium ions *in vivo*. Phosphorylase kinase activity would therefore approximately double for each 10% of phosphorylase b converted into phosphorylase a. This might provide a mechanism for preventing a decrease in the rate of phosphorylase a formation caused by depletion of the substrate phosphorylase b. The potential rate of this autocatalytic reaction *in vivo* is, however, perhaps fifty times slower than activation by the cyclic AMP-dependent kinase.

Activation of Phosphorylase Kinase by Proteolysis

The activation of phosphorylase kinase by the calcium-dependent proteinase in rabbit skeletal muscle described above, and the other proteinases such as trypsin, was described several years ago (Meyer *et al.*, 1964; Huston & Krebs, 1968). Prolonged storage of the highly purified enzyme at 4°C can also result in more than a 20-fold increase in activity at pH 6.8 (Cohen, 1973), particularly if the final step in the purification is omitted. Gel electrophoresis shows that the amount of α subunit is greatly decreased and that of the β subunit is greatly increased, suggesting that the preparation contains trace amounts of a proteinase capable of degrading the α subunit to a product of similar size to the β subunit, and that this limited proteolysis is responsible for the observed activation.

Plate 4 shows the result of incubating phosphorylase kinase (1.0 mg/ml) with

EXPLANATION OF PLATE I

Polyacrylamide-gel electrophoresis of phosphorylase kinase in the presence of sodium dodecyl sulphate

Migration is from top to bottom (reproduced from Cohen, 1973).

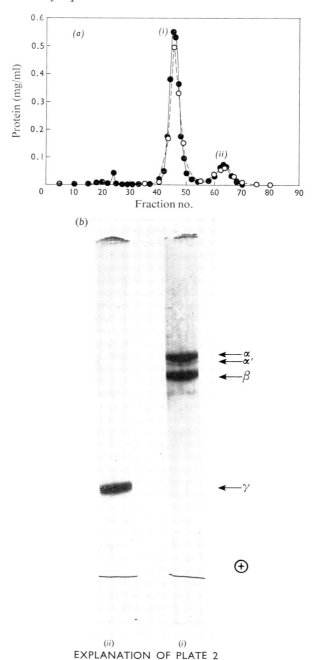

EXPLANATION OF PLATE 2

Fractionation of phosphorylase kinase

(a) Gel filtration of phosphorylase kinase on a column of Sephadex G-200 (150cm × 1.2cm) in the presence of sodium dodecyl sulphate. ●, Protein estimated from E_{280} assuming $E_{280}^{1\%} = 12.4$; ○, protein determined by the procedure of Lowry *et al.* (1951); fraction vol., 1ml. (b) Gel electrophoresis in the presence of sodium dodecyl sulphate of the pooled peaks (i) and (ii) from (a). Migration is from top to bottom (reproduced from Cohen, 1973).

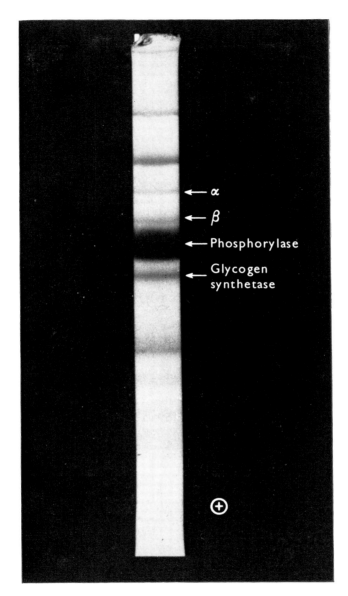

EXPLANATION OF PLATE 3

Polyacrylamide-gel electrophoresis of the glycogen particles in the presence of sodium dodecyl sulphate

Migration is from top to bottom (P. Cohen, unpublished work).

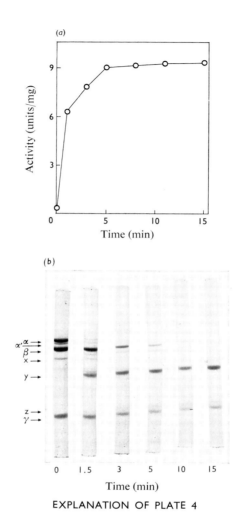

EXPLANATION OF PLATE 4

Effect of trypsin on the structure and activity of phosphorylase kinase

The enzyme (1.0 mg/ml) was incubated with trypsin (2 μg/ml) in buffer A at 25°C. (a) Increase in enzyme activity with time. (b) Alteration of polyacrylamide-gel pattern with time. Component X is a trace contaminant often present in phosphorylase kinase preparations. It may be eliminated by DEAE-cellulose chromatography (Fig. 1) (reproduced from Cohen, 1973).

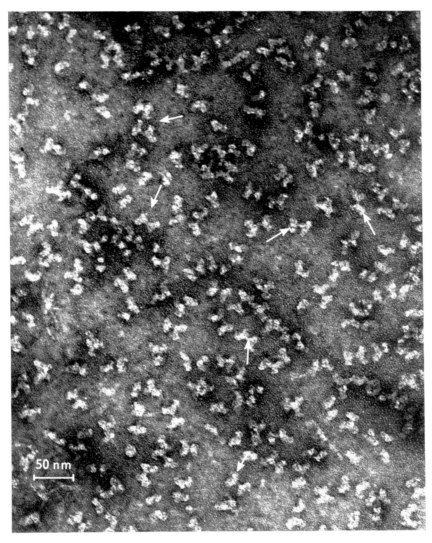

EXPLANATION OF PLATE 5

Electron micrographs of phosphorylase kinase

The enzyme is negatively stained with uranyl formate (M. T. Davison & P. Cohen, unpublished work). The arrows indicate some typical molecules.

trypsin (2μg/ml) at pH 6.8. Almost 100-fold activation was obtained within 5 min at pH 6.8. After 1 min, activation was about 80% complete, the α subunit had almost disappeared, and a new component, labelled 'Y', had formed. Thus activation by trypsin, like activation after prolonged storage, appears to result from a limited cleavage of the α subunit.

Between 1 and 15 min activity remained constant, although further peptide bonds were cleaved. The β subunit gradually disappeared, component Y increased in intensity and other bands, e.g. Z, appeared. The time-course of proteolysis suggested Y was an intermediate of the degradation of both the α and β subunits. Together with the observation that activation after prolonged storage arose from cleavage of the α subunit to a form migrating with the β subunit, the results are suggestive of a structural relationship between the α and β subunits. The γ subunit, which was resistant to phosphorylation, also appeared to be unaffected by proteolysis.

Specificity of Phosphorylase Kinase

Phosphorylase kinase activates phosphorylase by phosphorylation of a unique serine residue located 14 residues from the N-terminus of the protein, which contains 865 amino acid residues (Fischer et al., 1972). The sequences of dogfish and rabbit muscle phosphorylases in this region are shown in Fig. 8, and are characterized by a large number of basic amino acids. The linear sequence around the phosphoserine is thought to be an important recognition site for both phosphorylase kinase and phosphorylase phosphatase. Thus phosphorylase phosphatase will dephosphorylate the peptides comprising residues 5–18 or 11–16 at 2–3% of the rate at which it acts on phosphorylase a, but will not dephosphorylate phosphoserine and other phosphopeptides (Graves et al., 1960; Krebs et al., 1964). Phosphorylase kinase will phosphorylate the tetradecapeptide comprising residues 5 to 18, at a low rate, but only serine-14 and not serine-5 is phosphorylated (Tessmer & Graves, 1973). Carboxypeptidase B treatment of peptide 11–16 abolishes the action of phosphorylase phosphatase on the peptide, suggesting that arginine-16 plays an important role in the recognition of serine-14 by the phosphatase (Graves et al., 1960). Arginine methyl ester is a competitive inhibitor of phosphorylase phosphatase (Keller & Fried, 1955) and phosphorylase kinase (Tessmer & Graves, 1973).

Phosphorylase kinase will not phosphorylate any enzyme in the glycolytic pathway at an appreciable rate other than phosphorylase (Table 2), and, until recently, the only known physiological substrate for phosphorylase kinase other than phosphorylase was phosphorylase kinase itself. The enzyme has now

```
           1   2   3   4   5   6   7   8   9  10  11  12  13  14  15  16  17  18
                                                            *
(a) N-Acetyl-Ser-Arg-Pro-Leu-Ser-Asp-Gln-Glu-Lys-Arg-Lys-Gln-Ile-Ser-Val-Arg-Gly-Leu
                                                            *
(b) N-Acetyl-Ser-Lys-Pro-Lys-Ser-Asp-Met-Glu-Arg-Arg-Lys-Gln-Ile-Ser-Val-Arg-Gly-Leu
```

* phosphorylated by phosphorylase kinase.

Fig. 9. *N-Terminal sequences of rabbit (a) and dogfish (b) skeletal muscle phosphorylases* (Fischer et al., 1972)

been implicated in the phosphorylation of several muscle proteins directly concerned in the contractile process. Phosphorylase kinase was reported to phosphorylate the inhibitor component of troponin at rates not far below those obtained with phosphorylase (Stull *et al.*, 1972) and the dephosphorylation of this protein was catalysed by phosphorylase phosphatase (England *et al.*, 1972). However, Perry & Cole (1973) reported that the T component of troponin was phosphorylated by phosphorylase kinase, whereas the I component was phosphorylated at a slower rate. The cyclic AMP-dependent protein kinase has also been reported to phosphorylate both the I component of troponin (Bailey & Villar-Palasi, 1971) and the T component (Pratje & Heilmeyer, 1972) and also actin (Pratje & Heilmeyer, 1972). One of the 'light chains' associated with the head region of myosin may also be extracted in both a phosphorylated and dephosphorylated form (Perrie *et al.*, 1972). Although no functional differences between the phosphorylated and dephosphorylated forms of these proteins have been reported yet, it is possible that the phosphorylation of troponin, the protein that confers the calcium sensitivity on actomyosin ATPase (Ebashi *et al.*, 1969) could, for instance, alter the concentration of calcium ions at which the contractile process is activated. This suggests the possibility of a further role for phosphorylase kinase in the regulation of muscle metabolism through calcium ions. As the cyclic AMP-dependent protein kinase and phosphorylase kinase are also linked to adrenaline action, phosphorylation by either protein could provide the molecular basis for the positive inotropic effect of this catecholamine on skeletal muscle and heart muscle (Pratje & Heilmeyer, 1972; Stull *et al.*, 1972).

Regulation of Phosphorylase Kinase *in vivo*

The activities of the different forms of phosphorylase kinase in the standard assay are given in Table 3. The forms are named by analogy with the system adopted for its substrate phosphorylase, with 'b' representing the calcium-free

Table 2. *Phosphorylation of glycolytic enzymes by phosphorylase kinase*

Enzymes were from rabbit skeletal muscle except fructose 1,6-diphosphatase from rabbit liver. Experiments were carried out at 25°C in Tris–HCl, $I = 0.01$, pH 8.0 with 5 mM-magnesium acetate–1 mM-ATP (P. Cohen, unpublished work).

Enzyme	Relative activity
Phosphorylase b	100
Phosphoglucomutase	0.3
Phosphofructokinase	0.1
Aldolase	0.5
Triose phosphate isomerase	0.2
Triose phosphate dehydrogenase	0.5
Phosphoglycerate kinase	0.1
Phosphoglyceromutase	0.2
Enolase	0.5
Pyruvate kinase	0.1
Lactate dehydrogenase	0.1
Creatine kinase	0.2
Adenylate kinase	0.1
Fructose diphosphatase	0.1

dephosphorylated inactive enzyme and 'a' the dephosphorylated calcium-saturated active enzyme. The phosphorylated, calcium-saturated enzyme is termed 'sa' (super active at pH 6.8) and the phosphorylated, calcium-free inactive enzyme, 'sb'. The corresponding forms treated with trypsin, as in Plate 4, are termed b', a' and sa'.

Phosphorylase kinase a is the form isolated from rabbit skeletal muscle and has a pH 6.8/pH 8.2 activity ratio of 0.01–0.02 (Table 4). It is possible to estimate from the pH 6.8 activity of the a form, that the half-time for phosphorylase a formation in vivo after the release of calcium ions, would be 15 s (pH 6.8/pH 8.2 = 0.02) or 30 s (pH 6.8/pH 8.2 = 0.01). These times are considerably greater than the 1 s required to achieve 50% phosphorylase a formation when frog sartorius muscle is stimulated isometrically (Danforth et al., 1962). The time of 15–30 s might be an underestimate because phosphorylase kinase and phosphorylase are bound to glycogen and are therefore effectively at higher concentrations, whereas a uniform distribution was assumed in the above calculation. In addition, the half-time for phosphorylase a formation might be lowered by autocatalytic activation of phosphorylase kinase (Fig. 7), a reaction stimulated 2–3-fold by glycogen (Delange et al., 1968). The possible influence of other metabolites such as inorganic phosphate, reported to be an activator of insect flight muscle phosphorylase kinase (Hansford & Sacktor, 1970) has also not been considered. On the other hand, no allowance has been made for phosphorylase phosphatase which, if it were active, would decrease the net rate of phosphorylase a formation. (The corresponding calculation for phosphorylase kinase sa yields a half-time for phosphorylase a formation of 0.6 s.)

The above paragraph has assumed that the form of phosphorylase kinase which is isolated is type a, the undegraded and totally dephosphorylated enzyme. However two further possibilities must be considered.

(1) The pH 6.8 activity of the isolated enzyme may be produced by a slight proteolysis of the α subunit, and only a 1% degradation of this subunit during the isolation procedure would entirely account for the pH 6.8 activity of phosphorylase kinase α (Table 3). Indeed, the existence of the α' component (Plate 1) suggests that proteolysis of the enzyme has not been entirely avoided.

(2) Phosphorylase kinase 'a' might be contaminated with trace amounts of sa and again only a 2–3% contamination would entirely account for the observed activity.

Neither of the above possibilities has yet been ruled out, so that it is impossible

Table 3. *Activities of different forms of phosphorylase kinase*

The table is reproduced from Cohen (1973).

Enzyme form	Activity (units/mg)		Activity ratio pH 6.8/pH 8.2
	pH 6.8	pH 8.2	
b, b', sb	0	0	
a	0.11	9	0.012
sa	4.5	12.5	0.36
a'	10	16	0.67
sa'	10	16	0.67

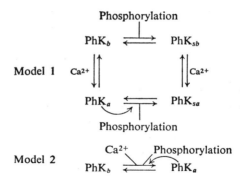

Scheme 2. *Alternative models for the regulation of phosphorylase kinase (PhK) by calcium ions and phosphorylation*
Evidence for each model is given in the text.

to distinguish between the two models shown in Scheme 2. Model 1 assumes that the dephosphorylated enzyme possesses activity in the presence of calcium and that there are therefore two active forms *a* and *sa*. Model 2 assumes that the slight activity observed in the purified enzyme at pH 6.8 is due to trace contamination with *a'* and/or *sa* and that there is just a single active species '*a*' for which both calcium ions and phosphorylation are required to form the active enzyme. If this latter model is correct, it is most likely that phosphorylase kinase would always be phosphorylated to a very small extent *in vivo*, even in resting muscle. In both models no phosphorylase *a* could be formed in the absence of calcium and the rate of phosphorylase *a* formation in the presence of calcium would be dependent on the degree of phosphorylation of the protein.

There is, however, one further possibility: that the muscle cell always contains trace amounts of the *a'* species. Phosphorylase kinase is presumably broken down and resynthesized with a finite half-life like other enzymes. Since all known proteinases appear to cause a similar activation of the enzyme, it is possible that at any given time skeletal muscle will contain a little of the *a'* form. If this is so, it would result in a model different from both 1 and 2 (Scheme 2), in which calcium ions could achieve a phosphorylase *b*-into-*a* conversion in the absence of any phosphorylation of phosphorylase kinase, but the activity would be entirely due to the trace amounts of the degraded form *a'* present in the cell!

A further role for the proteolytic activation of phosphorylase kinase after cell death has also been suggested (Cohen, 1973). Thus rupture of the lysosomes with ensuing release of its proteases should lead to a rapid activation of phosphorylase kinase and degradation of glycogen to small metabolites, which can then be re-absorbed and used by other cells (Cohen, 1973).

Regulation of Glycogen Breakdown and Synthesis

The role of phosphorylase kinase in the nervous and hormonal control of glycogen metabolism is shown in Scheme 3, in which the simple scheme, model 2 (Scheme 2), is used. Several general observations can be made.

The primary nervous and hormonal controls appear to be exerted on the two kinases in the Scheme 3, through calcium ions and cyclic AMP. The regulation of the phosphatases is not at all well understood, but the evidence suggests that the regulation of these activities may be exerted through conformational changes in their substrates rather than by changes in the conformation of the phosphatases themselves. Thus the dephosphorylation of glycogen synthetase is inhibited by the binding of glycogen to glycogen synthetase (Larner *et al.*, 1968), β subunit dephosphorylation of phosphorylase kinase is accelerated by α subunit phosphorylation as described above, and phosphorylase phosphatase is inhibited by the binding of AMP to its substrate phosphorylase *a* (Fischer *et al.*, 1971). The possibility of a second mechanism for regulating phosphorylase phosphatase has been suggested from studies with rabbit muscle glycogen particles (Haschke *et al.*, 1970), but the molecular basis of this effect has not been elucidated.

Scheme 3 suggests that the protein kinase and protein phosphatase provide a mechanism for the synchronous control of glycogen breakdown and synthesis. In fact, these two events are unlikely to occur at the same time. In resting muscle when large stores of glycogen have accumulated, glycogen synthetase is almost totally in the phosphorylated inactive state even in the absence of hormonal stimulation, whereas phosphorylase and phosphorylase kinase are essentially

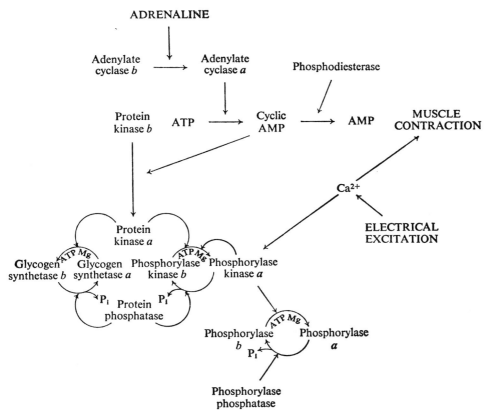

Scheme 3. *Nervous and hormonal regulation of glycogen metabolism*
b, Inactive enzyme; *a*, active enzyme.

dephosphorylated. The implied inhibition of glycogen synthetase phosphatase is probably caused by the feedback inhibition of this enzyme by glycogen referred to above (Larner et al., 1968). Thus although identical kinases and phosphatases appear to be used in the regulation of glycogen synthetase and phosphorylase kinase action, the control of these enzymes may be quite different if they are controlled through conformational changes in their enzyme substrates.

The potential activities of protein kinase, protein phosphatase, phosphorylase kinase and phosphorylase phosphatase in rabbit skeletal muscle are high and the optimal conditions measured *in vitro* show that they are each capable of acting *in vivo* on their respective substrates with a half-time of a few seconds at most. The molar proportions of phosphorylase, phosphorylase kinase and cyclic AMP-dependent protein kinase in rabbit skeletal muscle are approximately 240:10:1 respectively, taking the subunit molecular weight of phosphorylase as 100000 (Cohen et al., 1971), and the minimum active unit of phosphorylase kinase as 320000 ($\alpha\beta\gamma$) (Cohen, 1973) and cyclic AMP-dependent protein kinase as 123000 (Corbin et al., 1972). Phosphorylase phosphatase has been purified 300–700-fold (Fischer et al., 1971; England et al., 1972) from rabbit skeletal muscle and the active unit may have a molecular weight of about 30000 as judged by gel-filtration experiments (Fischer et al., 1971). Therefore phosphorylase phosphatase may still be present in a 1:1 molar ratio with phosphorylase kinase ($\alpha\beta\gamma$), since at least 1200–1500 enrichment from muscle extracts would be required for the complete purification.

Glycogenolysis in Mice Lacking Phosphorylase Kinase Activity

Despite the important role of phosphorylase kinase in the control of glycogenolysis, the enzyme does not appear to be essential to life, since I strain mice, which completely lack phosphorylase kinase activity, are relatively normal (Lyon & Porter, 1963). They are also able to break down glycogen after electrical excitation, although more slowly and after a lag period, compared with normal mice (Danforth & Lyon, 1964). The inheritance of the deficiency shows that it is determined by the X-chromosome (Lyon, 1970), and immunological evidence suggests the mice possess a totally inactive phosphorylase kinase, present in similar concentrations to the normal enzyme and giving an identical cross-reaction against antibody prepared to the pure rabbit skeletal muscle enzyme (Cohen & Cohen, 1973).

I strain mice are probably able to degrade glycogen by utilizing the alternative mechanism of phosphorylase *b* activation, by AMP (Scheme 1). The breakdown of glycogen should then be dependent on a rise in AMP/ATP ratio but unresponsive to immediate electrical or hormonal stimulation. The activation of phosphorylase *b* by AMP may therefore constitute an emergency mechanism for achieving glycogen breakdown should phosphorylase *a* formation fail to occur. Under normal conditions it is also possible that it provides a mechanism for prolonging glycogenolysis after electrical excitation ceases. Glycogenolysis would presumably continue until the creatine phosphate and ATP/AMP ratios return to resting muscle concentrations. The inhibition of phosphorylase phosphatase by AMP (Fischer et al., 1971) and phospho-dephospho hybrids of phosphorylase *b* and phosphorylase *a* (Hurd et al., 1966) may serve similar functions.

Electron Microscopy of Phosphorylase Kinase

Phosphorylase kinase must interact with at least three other enzymes of glycogen metabolism, phosphorylase, cyclic AMP-dependent protein kinase and phosphorylase kinase phosphatase. It also interacts with Mg–ATP, calcium ions and glycogen. Modification of the α subunit by proteolysis or phosphorylation of the β subunit each produce a large increase in activity, suggesting that these two chains function at least partially to hold the enzyme in an inactive conformation. The cyclic AMP-dependent protein kinase and phosphorylase kinase phosphatase must interact with both the site phosphorylated on the β subunit responsible for the activation, and the site phosphorylated on the α subunit, which may be important in regulating the dephosphorylation reaction. Phosphorylase b and ATP–Mg^{2+} must presumably bind to the subunit carrying the active site of the enzyme. However, the location of the active site and the calcium-binding site are not known. It is tempting to speculate that the γ subunit, for which no function has yet been found, is the catalytic subunit, and some evidence that this may be so has recently been reported (Graves et al., 1973). It was found that after extensive tryptic attack, phosphorylase kinase could be dissociated to an active 6S species in the presence of ATP, at 5°C and low protein concentration. However, although this species was greatly enriched in the γ subunit, it is still possible that the active site is located on fragments of the α and/or β units still present in this preparation. If the γ component is the catalytic subunit, it implies that 85% of the molecule ($\alpha+\beta$) is playing a regulatory rather than a catalytic role.

Plate 5 shows an electron micrograph of phosphorylase kinase. The molecules, some of which are indicated by arrows (Plate 5), typically show two large sections joined by a narrow interconnecting bridge. Each large section appears to be divided into two subsections and the bridge between the two large sections occurs near the junction of the subsections. The dimensions of each large section, approximately $20\,nm \times 8\,nm$, suggest they may each be composed of two α and two β subunits. The interconnecting bridge, approximately $7\,nm \times 3\,nm$, may represent the γ subunits. The interesting feature of this molecule are the holes in the structure that the 'H' shape creates. These are large enough to accommodate both phosphorylase b and cyclic AMP-dependent protein kinase, and it would be logical if both these two enzymes were able to interact simultaneously with phosphorylase kinase. It is also tempting to speculate further that in the inactive enzyme, the α and β subunits prevent the interaction of phosphorylase with the central catalytic γ subunit. Activation by phosphorylation or proteolysis of the β and α subunits respectively might cause a small displacement of the large sections to allow the interaction of phosphorylase b and the γ subunit to occur.

Although the discussion given above is at present only speculative, it is clear that phosphorylase kinase which interacts with the maximum number of components in the glycogenolytic sequence offers the best starting point from which to attempt a reconstruction of the whole system from the isolated components.

I am greatly indebted to Mrs. Carol Taylor for her excellent assistance. This work was supported by a grant from the Science Research Council (London, U.K.).

References

Ashby, C. D. & Walsh, D. A. (1972) *J. Biol. Chem.* **247**, 6637–6642
Barrett, A. J. & Dingle, J. T. (1972) *Biochem. J.* **127**, 439–441
Bailey, C. & Villar-Palasi, C. (1971) *Fed. Proc. Fed. Amer. Soc. Exp. Biol.* **30**, 1147
Brostrom, C. O., Hunkeler, F. L. & Krebs, E. G. (1971a) *J. Biol. Chem.* **246**, 1961–1967
Brostrom, C. O., Corbin, J. D. & Krebs, E. G. (1971b) *Fed. Proc. Fed. Amer. Soc. Exp. Biol.* **30**, 1089
Brown, D. H. & Illingworth, B. (1962) *Proc. Nat. Acad. Sci. U.S.* **48**, 1783–1787
Cohen, P. (1972) *Biochem. J.* **130**, 5p–6p
Cohen, P. (1973) *Eur. J. Biochem.* **34**, 1–14
Cohen, P. & Antoniw, J. F. (1973) *FEBS Lett.* **34**, 43–47
Cohen, P. T. W. & Cohen, P. (1973) *FEBS Lett.* **29**, 113–118
Cohen, P., Duewer, T. & Fischer, E. H. (1971) *Biochemistry* **10**, 2683–2694
Corbin, J. D., Brostrom, C. O., King, C. A. & Krebs, E. G. (1972) *J. Biol. Chem.* **247**, 7790–7798
Cori, C. F. & Cori, G. T. (1936) *Proc. Soc. Exp. Biol. Med.* **34**, 702–705
Cori, C. F. & Larner, J. (1951) *J. Biol. Chem.* **188**, 17–29
Danforth, W. H. & Lyon, J. B. (1964) *J. Biol. Chem.* **239**, 4047–4050
Danforth, W. H., Helmreich, E. & Cori, C. F. (1962) *Proc. Nat. Acad. Sci. U.S.* **48**, 1191–1199
Delange, R. J., Kemp, R. G., Riley, W. D., Cooper, R. A. & Krebs, E. G. (1968) *J. Biol. Chem.* **243**, 2200–2208
Drummond, G. I., Harwood, J. P. & Powell, C. A. (1969) *J. Biol. Chem.* **244**, 4235–4240
Ebashi, S., Endo, M. & Ohtsuki, I. (1969) *Quart. Rev. Biophys.* **2**, 351–383
England, P. J., Stull, J. T. & Krebs, E. G. (1972) *J. Biol. Chem.* **247**, 5275–5277
Fischer, E. H., Heilmeyer, L. M. G. & Haschke, R. H. (1971) *Curr. Top. Cell. Regul.* **4**, 211–251
Fischer, E. H., Cohen, P., Fosset, M., Muir, L. W. & Saari, J. C. (1972) in *Metabolic Interconversions of Enzymes* P11 (Weiland, O., Helmreich, E. & Holzer, H., eds.), pp. 11–28, Springer-Verlag, Heidelberg
Graves, D. J., Fischer, E. H. & Krebs, E. G. (1960) *J. Biol. Chem.* **235**, 805–809
Graves, D. J., Hayakawa, T., Horwitz, R. A., Beckman, E. & Krebs, E. G. (1973) *Biochemistry* **12**, 580–585
Hansford, R. G. & Sacktor, B. (1970) *FEBS Lett.* **7**, 183–187
Haschke, R. H., Heilmeyer, L. M. G., Meyer, F. & Fischer, E. H. (1970) *J. Biol. Chem.* **245**, 6657–6662
Hayakawa, T., Perkins, J. P., Walsh, D. A. & Krebs, E. G. (1973a) *Biochemistry* **12**, 567–573
Hayakawa, T., Perkins, J. P. & Krebs, E. G. (1973b) *Biochemistry* **12**, 574–580
Heilmeyer, L. M. G., Meyer, F., Haschke, R. H. & Fischer, E. H. (1970) *J. Biol. Chem.* **245**, 6649–6656
Hurd, S. S., Teller, D. C. & Fischer, E. H. (1966) *Biochem. Biophys. Res. Commun.* **24**, 79–84
Huston, R. B. & Krebs, E. G. (1968) *Biochemistry* **7**, 2116–2122
Keller, P. J. & Fried, M. (1955) *J. Biol. Chem.* **214**, 143–148
Krebs, E. G. (1972) *Curr. Top. Cell. Regul.* **5**, 99–133
Krebs, E. G. & Fischer, E. H. (1956) *Biochim. Biophys. Acta* **20**, 150
Krebs, E. G., Love, D. S., Bratvold, G. E., Trayser, K. A., Mayer, W. L. & Fischer, E. H. (1964) *Biochemistry* **3**, 1022–1033
Larner, J., Villar-Palasi, C., Goldberg, N. D., Bishop, J. S., Huijing, F., Wenger, J. I., Sasko, H. & Brown, N. B. (1968) *Advan. Enzyme Regul.* **6**, 409–423
Lowry, O. H., Rosebrough, N. J., Farr, A. L. & Randall, R. J. (1951) *J. Biol. Chem.* **193**, 265–275
Lyon, J. B. (1970) *Biochem. Genet.* **4**, 169–185
Lyon, J. B. & Porter, J. (1963) *J. Biol. Chem.* **238**, 1–11
Meyer, F., Heilmeyer, L. M. G., Haschke, R. H. & Fischer, E. H. (1970) *J. Biol. Chem.* **245**, 6642–6648
Meyer, S. E. & Krebs, E. G. (1970) *J. Biol. Chem.* **245**, 3153–3160
Meyer, W. L., Fischer, E. H. & Krebs, E. G. (1964) *Biochemistry* **3**, 1033–1039
Ozawa, E., Hosoi, K. & Ebashi, S. (1967) *J. Biochem. (Tokyo)* **61**, 531–533
Perrie, W. T., Smillie, L. B. & Perry, S. V. (1972) *Biochem. J.* **128**, 105p–106p
Perry, S. V. & Cole, H. A. (1973) *Biochem. J.* **131**, 425–428
Posner, J. B., Stern, R. & Krebs, E. G. (1965) *J. Biol. Chem.* **240**, 982–985
Pratje, E. & Heilmeyer, L. M. G. (1972) *FEBS Lett.* **27**, 89–93
Rall, T. W., Sutherland, E. W. & Wosilait, W. D. (1956) *J. Biol. Chem.* **218**, 483–495
Rall, T. W., Sutherland, E. W. & Berther, J. (1957) *J. Biol. Chem.* **224**, 463–475
Reimann, E. M., Walsh, D. A. & Krebs, E. G. (1971) *J. Biol. Chem.* **246**, 1986–1995
Riley, W. D., Delange, R. J., Bratvold, G. E. & Krebs, E. G. (1968) *J. Biol. Chem.* **243**, 2209–2215

Rubin, C. S., Erlichman, J. & Rosen, O. M. (1972) *J. Biol. Chem.* **247**, 36–44
Soderling, T. R., Hickinbottom, J. P., Reimann, E. M., Hunkeler, F. L., Walsh, D. A. & Krebs, E. G. (1970) *J. Biol. Chem.* **245**, 6317–6328
Stull, J. T., Brostrom, C. O. & Krebs, E. G. (1972) *J. Biol. Chem.* **247**, 5272–5274
Tessmer, G. & Graves, D. J. (1973) *Biochem. Biophys. Res. Commun.* **50**, 1–7
Walsh, D. A., Perkins, J. P. & Krebs, E. G. (1968) *J. Biol. Chem.* **243**, 3763–3765
Walsh, D. A., Perkins, J. P., Brostrom, C. O., Ho, E. S. & Krebs, E. G. (1971a) *J. Biol. Chem.* **246**, 1968–1976
Walsh, D. A., Ashby, C. D., Gonzalez, C., Calkins, D., Fischer, E. H. & Krebs, E. G. (1971b) *J. Biol. Chem.* **246**, 1977–1985
Zieve, F. J. & Glinsmann, W. H. (1973) *Biochem. Biophys. Res. Commun.* **50**, 872–878

Calcium Ions and the Regulation of Pyruvate Dehydrogenase

By PHILIP J. RANDLE, RICHARD M. DENTON, HELEN T. PASK and DAVID L. SEVERSON

Department of Biochemistry, University of Bristol, Medical School, University Walk, Bristol BS8 1TD, U.K.

Synopsis

Mammalian pyruvate dehydrogenases are inactivated by phosphorylation with ATPMg^{2-} catalysed by pyruvate dehydrogenase kinase. Dephosphorylation and reactivation is catalysed by pyruvate dehydrogenase phosphate phosphatase. The phosphatase from pig heart, pig kidney cortex, rat heart and rat epididymal adipose tissue has been shown to require both Mg^{2+} and Ca^{2+} for activity. With pig heart pyruvate dehydrogenase phosphate and the pig heart phosphatase the K_m for Mg^{2+} is approximately 1 mM. With calcium–EGTA [ethanedioxybis(ethylamine)tetra-acetate] buffers the K_m for Ca^{2+} is approx. 0.7 μM. The effect of Ca^{2+} is to lower the K_m of the phosphatase for pyruvate dehydrogenase phosphate (from 30 μM with 10 mM-EGTA to 1.6 μM with 10 mM-EGTA + 9.74 mM-CaCl$_2$, computed Ca^{2+} 10 μM). In fat-cell mitochondria, as isolated, the Ca^{2+} concentration is apparently sufficient to activate the phosphatase. It is concluded that Ca^{2+} is a cofactor for the phosphatase but that a role for Ca^{2+} as a regulator of the phosphatase has yet to be firmly established.

Introduction

Pyruvate dehydrogenase catalyses the conversion of pyruvate into acetyl-CoA by an essentially irreversible reaction. This reaction is of critical importance in the economy of mammals, because as a result the potential for resynthesis of glucose from products of its catabolism is lost. It serves both bioenergetic and biosynthetic functions. Pyruvate oxidation is an important source of ATP formation in heart muscle, kidney and brain and to a lesser extent in skeletal muscles. In liver and adipose tissue the reaction provides acetyl-CoA for the biosynthesis of fatty acids, triglyceride and cholesterol from glucose and other sugars. Regulation of pyruvate dehydrogenase is thus important in control of the tricarboxylate cycle, in control of the contributions of pyruvate, fatty acids and ketone bodies to respiration and in control of lipid biosynthesis.

Mammalian pyruvate dehydrogenases are subject to end-product inhibition by high ratios of [acetyl-CoA]/[CoA] and [NADH]/[NAD$^+$]. This may contribute to regulation of pyruvate oxidation in heart muscle by the oxidation of fatty acids and ketone bodies (Garland & Randle, 1964). As first shown by Linn *et al.* (1969a) pyruvate dehydrogenase can exist in phosphorylated (inactive) and dephosphorylated (active) forms. Interconversion of these two forms is effected by a kinase and a phosphatase. In 1972 Denton *et al.*, Siess & Wieland and Pettit *et al.* published evidence that Ca^{2+} may activate the phosphatase thus suggesting the possibility of a role for this metal ion in the regulation of pyruvate dehydrogenase activity. Before discussing evidence on this point it may be helpful to summarize current knowledge of the pyruvate dehydrogenase complex and of its regulation and to describe factors that regulate pyruvate oxidation and pyruvate dehydrogenase activity in mammalian cells and mitochondria.

Component Enzymes of the Pyruvate Dehydrogenase Complex

The pyruvate dehydrogenase complex (referred to as pyruvate dehydrogenase, EC 1.2.4.1) catalyses the reactions shown below (Hucho et al., 1972).

(1) Pyruvate + thiamin-P-P $\xrightarrow{enz._1}$ CO_2 + hydroxyethyl-thiamin-P-P

(2) Hydroxyethyl-thiamin-P-P + lipoate $\underset{enz._1}{\rightleftharpoons}$ acetylhydrolipoate + thiamin-P-P

(3) Acetylhydrolipoate + CoA $\underset{enz._2}{\rightleftharpoons}$ acetyl-CoA + dihydrolipoate

(4) Dihydrolipoate + NAD^+ $\underset{enz._3}{\rightleftharpoons}$ lipoate + NADH + H^+

(5) $Enz._1$ + ATP $\xrightarrow{enz._4}$ $enz._1$-P + ADP

(6) $Enz._1$-P + H_2O $\xrightarrow{enz._5}$ $enz._1$ + P_i

In this scheme, the abbreviations are: $enz._1$, pyruvate decarboxylase; $enz._2$, dihydrolipoate acetyltransferase; $enz._3$, dihydrolipoate dehydrogenase; $enz._4$, pyruvate dehydrogenase kinase; $enz._5$, pyruvate dehydrogenase phosphate phosphatase; thiamin-P-P, thiamin pyrophosphate.

Inhibition of pyruvate dehydrogenase by high ratios of [acetyl-CoA]/[CoA] and of [NADH]/[NAD^+] may be due to acetylation and reduction of lipoate in the complex as a result of equilibrium considerations in reactions (3) and (4). Control by phosphorylation–dephosphorylation proceeds according to reactions (5) and (6) (Linn et al., 1969a,b, 1972; Barrera et al., 1972; Hucho et al., 1972; Roche & Reed, 1972).

Analyses of bovine heart and kidney complexes by Dr. Reed and his colleagues (for references see preceding paragraph) showed a molecular weight of approx. 10^7 for pyruvate dehydrogenase with a specific activity of approx. 10 units/mg. The complex contained as a core dehydrolipoyl acetyltransferase (mol.wt. 3.1×10^6; 60 polypeptide chains) to which were attached 30 molecules of pyruvate decarboxylase ($\alpha_2\beta_2$, see below; mol.wt. 1.54×10^5), 10–12 molecules of dihydrolipoate dehydrogenase and 5 molecules of pyruvate dehydrogenase kinase. These components were relatively tightly bound and not separated in the ultracentrifuge; separation was achieved by chromatography on Sepharose 4B. The phosphatase was less tightly bound and did not sediment with the complex in the ultracentrifuge; there were approx. 5 molecules of phosphatase (mol.wt. approx. 1.00×10^5) associated with each complex. Further studies indicated that the pyruvate decarboxylase component contains two non-identical polypeptide chains (α, mol.wt., 4.1×10^4; β, mol.wt., 3.6×10^4). Phosphorylation [reaction (5) in the scheme] results in incorporation of phosphate into the α chains of the decarboxylase (possibly 1 mol of phosphate/mol of α chain) and to specific loss of activity in reaction (1) of the scheme. This has suggested that reaction (1) may be catalysed by α chains and reaction (2) by β chains (Roche & Reed, 1972).

Effectors of Pyruvate Dehydrogenase Kinase; Factors Modulating Pyruvate Dehydrogenase Activity in Tissues and Mitochondria

Phosphorylation and inactivation of mammalian pyruvate dehydrogenase complexes is inhibited by ADP (competitive with ATP), by pyruvate (non-competitive with ATP) and by phosphate, pyrophosphate and thiamin pyrophosphate (Linn et al., 1969a; Hucho et al., 1972; Martin et al., 1972; Randle & Denton, 1973). Phosphorylation is also inhibited by dichloroacetate (Whitehouse & Randle, 1973). If these agents are effective in vivo then it is to be expected that high ratios of [ATP]/[ADP] would inactivate pyruvate dehydrogenase and vice versa and that pyruvate, phosphate, pyrophosphate, thiamin pyrophosphate and dichloroacetate would activate pyruvate dehydrogenase provided that pyruvate dehydrogenase phosphate phosphatase is active. In fat-cell mitochondria or liver mitochondria, pyruvate dehydrogenase is inactivated on incubation with substrates for respiratory chain phosphorylation (2-oxoglutarate+malate, or succinate) and activated by the further inclusion of uncouplers or pyruvate (Martin et al., 1972; Portenhauser & Wieland, 1972). In rat epididymal fat-pads, pyruvate dehydrogenase is activated on incubation with uncouplers or pyruvate or glucose or fructose (which give rise to pyruvate by glycolysis) (Martin et al., 1972). Rat heart pyruvate dehydrogenase is substantially activated by perfusion with media containing dichloroacetate (Whitehouse & Randle, 1973).

In rat epididymal adipose tissue, pyruvate dehydrogenase is activated by insulin action and the effects of insulin are inhibited by adrenaline, adrenocorticotropic hormone or dibutyryl cyclic AMP. Ouabain, which reproduces many of the effects of insulin on adipose-tissue metabolism, also activates pyruvate dehydrogenase (Jungas & Taylor, 1972; Coore et al., 1971; Martin et al., 1972). In rat heart, pyruvate dehydrogenase activity is diminished by respiration of fatty acids, ketone bodies and acetate. In rat liver mitochondria, respiration of palmitoyl-L-carnitine leads to inactivation of pyruvate dehydrogenase (Wieland et al., 1971; Portenhauser & Wieland, 1972). The mechanism of the effects of hormones and ouabain in adipose tissue or of fatty acids and ketone bodies in perfused-heart or rat liver mitochondria is not known. Current evidence suggests that these effects cannot be explained by changes in tissue concentrations of pyruvate, ADP or ATP. Attempts to demonstrate effects of cyclic AMP on the activity of pyruvate dehydrogenase kinase or of the phosphatase with purified enzymes, mitochondrial extracts or intact fat-cell mitochondria have been unsuccessful. There is the possibility that modulation of phosphatase activity by Ca^{2+} concentration could contribute to the actions of hormones on pyruvate dehydrogenase activity in adipose tissue and to effects of fatty acids and ketone bodies in heart muscle.

The localization of pyruvate dehydrogenase in the mitochondrion is of some importance to its regulation. In particular it is important to know whether the kinase utilizes intramitochondrial ATP and whether the kinase and phosphatase are influenced solely by intramitochondrial effectors. When pig heart, pig kidney-cortex or rat fat-cell mitochondria are disrupted by repeated freezing and thawing approx. 30–50% of the pyruvate dehydrogenase activity is associated with the pellet on centrifuging at 15000g for 15min. Kinase and phosphatase activities are also present in this fraction. A substantial part of the pyruvate dehydrogenase may therefore be bound to mitochondrial membranes. The use of intramitochondrial ATP for phosphorylation of pyruvate dehydrogenase is suggested by the observation that oxidation of oxoglutarate+malate by fat-cell mitochondria leads to inactivation of the dehydrogenase in the absence of external ADP (Martin et al., 1972). As shown below, extramitochondrial Ca^{2+} does not appear to be required for phosphatase activity. The cellular localization of the action of other effectors has not been proved.

Discussion of the role of Ca^{2+} is concerned with the requirement of the phosphatase for this metal ion and with attempts to show that the ion may regulate pyruvate dehydrogenase in fat cells or fat-cell mitochondria.

Calcium Ion Concentration and Activity of Pyruvate Dehydrogenase Phosphate Phosphatase

Reagents

Pyruvate dehydrogenase phosphate phosphatase has been assayed by the release of $^{32}P_i$ from ^{32}P-labelled pyruvate dehydrogenase phosphate. Control experiments with the substrate preparations used in these studies have shown close correspondence between release of $^{32}P_i$ and pyruvate dehydrogenase activity. Pyruvate dehydrogenase was prepared from pig heart mitochondria by fractionation with polyethylene glycol (Linn et al., 1972) followed by precipitation of contaminating material (notably adenosine triphosphatase activity) at pH 6.1. Phosphatase was removed by sedimentation (three times) of pyruvate dehydrogenase at 150000g for 90min. This relatively crude preparation of phosphatase was used for some experiments. Pig-heart pyruvate dehydrogenase phosphate phosphatase was also partially purified by a modification of the method of Siess & Wieland (1972). Chromatography on DEAE-cellulose (by using elution with a linear gradient of 0–0.5M-KCl in 20mM-phosphate containing 1mM-$MgCl_2$, 2mM-2-mercaptoethanol, pH7) was used in place of batch treatment. Our preparations of phosphatase were unstable, losing up to 80% of activity with storage at −10°C for 2 weeks, or with freeze-drying, vacuum concentration or $(NH_4)_2SO_4$ precipitation. Because of this instability further purification has not been possible. The factors responsible for the development of this instability during purification have yet to be defined.

Effects of ethanedioxybis(ethylamine)tetra-acetate; requirements for Mg^{2+} and Ca^{2+}

Linn et al. (1969a,b) observed that pyruvate dehydrogenase phosphate phosphatases require Mg^{2+}. With an excess of Mg^{2+} (16mM) Denton et al. (1972) found that 10mM-EGTA [ethanedioxybis(ethylamine)tetra-acetate] markedly inhibits phosphatase activity with partially purified preparations of the pig heart enzyme and in extracts of rat fat-cell and pig kidney-cortex mitochondria. Similar inhibition by EGTA was observed by Siess & Wieland (1972). The use of calcium–EGTA buffers by Denton et al. (1972) indicated that the inhibitory effects of

EGTA are reversed by calcium and that the phosphatase is sensitive to Ca^{2+} in the range 0.01–10 μM. By the use of EDTA, Mg^{2+}, EGTA and calcium–EGTA buffers evidence was obtained for a dual requirement of pig heart phosphatase for Mg^{2+} and Ca^{2+}. With 1mM-EDTA (to remove contaminating Mg^{2+}) the enzyme was inactive and was not activated by the addition of calcium–EGTA buffers that yielded maximum activation in the presence of Mg^{2+}. With 10mM-EGTA (to remove contaminating Ca^{2+}) some activation was seen with 16mM-$MgCl_2$ but full activation required Mg^{2+} and calcium–EGTA. These findings led to the tentative conclusion of a dual requirement for Mg^{2+} and Ca^{2+}. The enzyme is active with Mg^{2+} in the absence of EGTA (30–100% of maximum activity) and it has been assumed that this is due to contamination of the reagents used with Ca^{2+}. Some reactivation in the presence of EGTA was seen with strontium and perhaps barium but the effects were much less marked than those of an equivalent concentration of calcium. These results have been given elsewhere (Denton *et al.*, 1972).

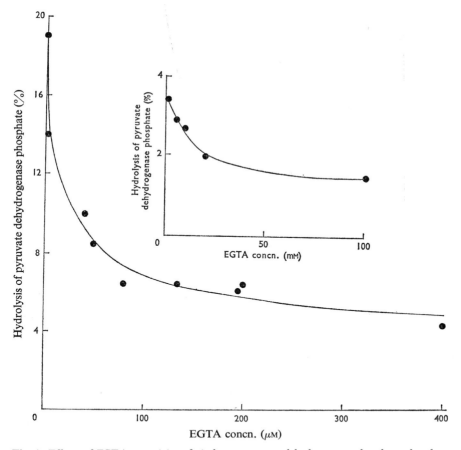

Fig. 1. *Effects of EGTA on activity of pig heart pyruvate dehydrogenase phosphate phosphatase*

The conditions of assay were as in Table 1 except that 85mM-$MgCl_2$ was used with 100mM-EGTA and 64mM-$MgCl_2$ with 50mM-EGTA. The percentage hydrolysis with 10mM-EGTA and 9.75mM-$CaCl_2$ (not shown) was 38%.

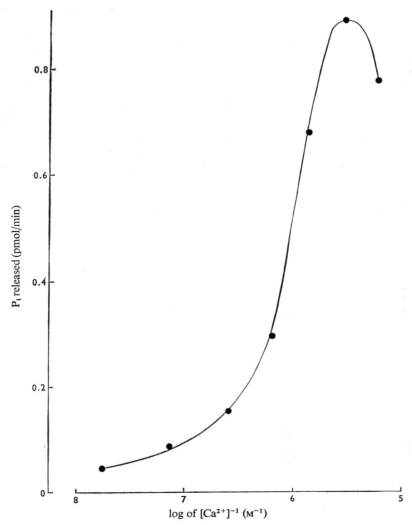

Fig. 2. *Effects of Ca^{2+} in calcium–EGTA buffers on activity of pig heart pyruvate dehydrogenase phosphate phosphatase*

The conditions of assay were as in Table 1 except that the concentration of $MgCl_2$ was adjusted to give a calculated Mg^{2+} concentration of approx. 4.5 mM (4.3–4.7 mM). The concentrations of Ca^{2+} and Mg^{2+} were calculated from the following dissociation constants: calcium–EGTA 1.31×10^{-7}; magnesium–EGTA, 1.88×10^{-2}; calcium phosphate, 4.1×10^{-3}; magnesium phosphate, 3.1×10^{-3} (assuming $HPO_4^{2-}/H_2PO_4^-$ at pH 7 is 0.69).

The K_m of the pig heart phosphatase for Mg^{2+} was approx. 1 mM in phosphate–mercaptoethanol, pH 7 (the pH optimum) containing 10 μM-$CaCl_2$. This is lower than values reported previously for phosphatases from a number of mammalian tissues. This difference is mainly due to corrections that we have applied for the binding of magnesium by phosphate.

The effects of EGTA concentration on the activity of the pig heart phosphatase are shown in Fig. 1. Significant inhibition required less than 16 μM-EGTA and

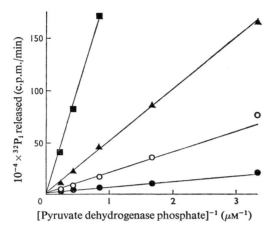

Fig. 3. *Effects of substrate concentration on the activity of pyruvate dehydrogenase phosphate phosphatase*

The conditions of assay were as given in Table 1 except that the concentration of Mg^{2+} was 20mM and that of Ca^{2+} was varied as follows: K_m values for pyruvate dehydrogenase phosphate phosphatase are given in parentheses. ■, Zero Ca^{2+} (30μM); ▲, 0.28μM-Ca^{2+} (12μM); ○, 1μM-Ca^{2+} (7.8μM); ●, 10μM-Ca^{2+} (1.6μM).

near-maximum inhibition required less than 400μM-EGTA. The effects of Ca^{2+} concentration (with calcium–EGTA buffers) are shown in Fig. 2; half-maximum activation was seen with approx. 0.7μM-Ca^{2+} (measurements in phosphate–mercaptoethanol at 4.5mM-Mg^{2+}). As shown in Fig. 3 progressive lowering of Ca^{2+} by the use of calcium–EGTA buffers results in a marked increase in the K_m of the phosphatase for pyruvate dehydrogenase phosphate. Fig. 3 has been drawn on the assumption that $V_{max.}$ is unchanged but this has only been shown statistically at the highest two concentrations of Ca^{2+}. Similar findings have been reported by Pettit *et al.* (1972) who suggested that the role of Ca^{2+} is to facilitate binding of the phosphatase to the dihydrolipoate acetyltransferase component of pyruvate dehydrogenase. We have no evidence bearing on this suggested mechanism at present. We have been unable to detect binding of ^{45}Ca to pyruvate dehydrogenase phosphate, but we have not been able to ascertain whether ^{45}Ca is bound by the phosphatase or by the complex formed with its substrate.

It has seemed to us necessary to attempt to show Ca^{2+} dependence by methods that do not involve EGTA or calcium–EGTA buffers. This has been explored by utilizing Chelex 100 resin to remove Ca^{2+} from buffers (phosphate–mercaptoethanol or 100mM-triethanolamine chloride, 2mM-mercaptoethanol, pH7). The use of ^{45}Ca showed that this treatment removes approx. 98–99% of the calcium. Preparations of phosphatase and of pyruvate dehydrogenase phosphate containing 1mM-EGTA were dialysed for 24–48h against 2–4 changes of calcium-depleted buffer. Phosphatase activity was then assayed with Spectrograde $MgSO_4$. This treatment led to a substantial decrease in the activity of the phosphatase, which in the best experiments was not significantly different from that seen with 10mM-EGTA (Table 1). Reactivation was readily achieved with calcium–EGTA (computed $[Ca^{2+}]$ 12μM). Reactivation was also achieved with analytical reagent-grade $CaCl_2$ in the absence of EGTA and significant stimulation was consistently

Table 1. *Effects of CaCl₂ on activity of pig heart pyruvate dehydrogenase phosphate phosphatase with reagents depleted of calcium*

Phosphate buffer (20 mM-potassium phosphate, 2 mM-2-mercaptoethanol, pH 7) or triethanolamine buffer (0.1 M-triethanolamine chloride, 2 mM-2-mercaptoethanol) was depleted of calcium by passage through a column of Chelex 100 resin. Pyruvate dehydrogenase phosphate phosphatase and ^{32}P-labelled pyruvate dehydrogenase phosphate containing 1 mM-EGTA were dialysed against calcium-depleted buffers. Phosphatase activity was assayed by the formation of ^{32}P$_i$ at 30 °C for 5 min in 20 μl of incubated sample containing 50×10^{-3} units of substrate (as units of pyruvate dehydrogenase activity) and sufficient phosphatase to give 10–20% hydrolysis. The results are expressed as means±S.E.M. The concentration of MgCl₂ was 32 mM. Calculated [Ca^{2+}] in calcium–EGTA buffer was 12 μM.

	Rate of hydrolysis (pmol of P$_i$/min) in:				
	Buffer	EGTA (10 mM)	5.5 μM-CaCl₂	500 μM-CaCl₂	EGTA (10 mM) + CaCl₂ (9.75 mM)
Phosphate	0.24±0.004	0.16±0.007	0.34±0.002	0.88±0.05	1.04±0.04
Triethanolamine	0.04±0.002	0	–	0.3 ±0.02*	0.54±0.04
	0.70±0.02	0.19±0.04	0.83±0.04	1.5 ±0.03	1.44±0.02

* The CaCl₂ concentration was 100 μM.

obtained with 5.5 μM-Ca^{2+}. However, much higher concentrations were required for complete reactivation (approx. 500 μM). This is some 50 times greater than the apparent requirement for Ca^{2+} in activation with calcium–EGTA buffers (approx. 12 μM). Representative data are in Fig. 4. There are a number of possible explanations for this discrepancy that have yet to be resolved. Adsorption of Ca^{2+} to plastic or glass might occur in the absence of EGTA buffering. Other metals present in the CaCl₂ used might antagonize the Ca^{2+} effect but be removed by EGTA. It is possible that calcium–EGTA may be a more active species than Ca^{2+} itself. It should also be emphasized that the calculated Ca^{2+} concentrations in calcium–

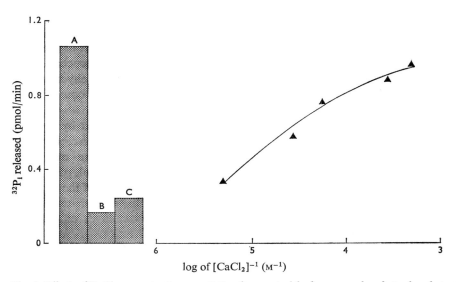

Fig. 4. *Effects of CaCl₂ concentration on activity of pyruvate dehydrogenase phosphate phosphatase*
The conditions of assay were as given in Table 1. A, calcium–EGTA (12 μM-Ca^{2+}). B, EGTA. C. low-[Ca^{2+}] buffer.

EGTA buffers are only approximate, because the published association constants were determined under conditions different from those utilized here.

The specificity of the metal requirements of mammalian pyruvate dehydrogenase phosphate phosphatases has not been exhaustively investigated. Hucho et al. (1972) observed that Mn^{2+} can replace Mg^{2+}. We have explored the ability of a number of divalent metals to substitute for Ca^{2+} by utilizing Ca^{2+}-depleted reagents (see above) and 16 mM-Mg^{2+}. At a concentration of 100 μM no stimulation was seen with Ba^{2+}, Zn^{2+}, Cu^{2+}, Co^{2+} or Fe^{2+}; 100 μM-Ca^{2+} produced a 4-fold stimulation and some stimulation (up to 2-fold) was seen with Sr^{2+}, La^{3+} and Mn^{2+}. It seems unlikely therefore that contamination of $CaCl_2$ with these metal ions could account for its stimulatory effects. It seems reasonable to conclude that the phosphatase requires calcium as a cofactor in addition to magnesium.

To date no other effectors of pig heart, bovine heart or bovine kidney phosphatases have been described. In studies with the partially purified pig heart enzyme with calcium–EGTA or EGTA we have seen no effects of ATP, ADP, AMP, GTP, cyclic GMP, cyclic CMP, cyclic UMP, cyclic dTMP, pyruvate, acetyl-CoA, CoA, oxaloacetate, citrate, 2-oxoglutarate, glutamate, succinate, malate or aspartate. No effects of cyclic AMP with or without protein kinase and ATP have been detected. Hucho et al. (1972) could detect little if any effect of acetyl-CoA, acetoacetyl-CoA, citrate, isocitrate, palmityl-CoA, phosphoenolpyruvate or cyclic AMP.

Is Calcium Involved in the Regulation of Pyruvate Dehydrogenase Phosphate Phosphatase *in vivo*?

Theoretical consideration

The rapidly exchangeable calcium in mitochondria in fat-cells is of the order of 20 nmol/g dry wt. of cells or approx. 4 μmol/ml of mitochondrial water (see below). The activity of the phosphatase in mitochondria is presumably dependent upon the concentration of Ca^{2+}, but this is not known. Mitochondria contain many different classes of compounds that are capable of binding calcium (e.g. nucleoside triphosphates, phosphate, phospholipids, citrate). The system may therefore be regarded as a complex buffer of ionizable calcium. Mitochondrial [Ca^{2+}] could therefore be varied by fluxes of calcium across the inner membrane changing total calcium; by changes in the concentrations of binding substances; and possibly by changes in the affinity for calcium of some of the binding substances. We have no evidence on these points at present.

Experiments with EGTA and calcium–EGTA buffers suggest that Ca^{2+} may lower the K_m of the phosphatase for its substrate. This consideration suggests that Ca^{2+} may be most effective *in vivo* at low concentrations of pyruvate dehydrogenase phosphate, i.e. at high concentrations of pyruvate dehydrogenase. The concentration of pyruvate dehydrogenase phosphate in rat hearts perfused with 16.5 mM-glucose and insulin is 20 units/g dry wt. (expressed as units of pyruvate dehydrogenase) or approx. 40 units/ml of mitochondrial water. Since part of this may be bound to mitochondrial membranes and since we have no kinetic results for the rat heart phosphatase with the rat heart substrate, speculations concerning the significance of this concentration would be premature. It may be noted, however, that the pig heart phosphatase has a K_m for its substrate of 50 units/ml in 10 mM-EGTA and a K_m of 4 units/ml in 10 mM-EGTA, 9.75 mM-CaCl$_2$ (computed to give 12 μM-Ca^{2+}).

Does Extramitochondrial Calcium Concentration Influence Activation of Pyruvate Dehydrogenase?

Experiments with mitochondria

Pyruvate, an inhibitor of pyruvate dehydrogenase kinase, increases the activity of pyruvate dehydrogenase in fat-cell mitochondria *in vitro* (Martin *et al.*, 1972). Hence the rate and extent of activation of the dehydrogenase by pyruvate should be diminished in the absence of external Ca^{2+} if this is required for phosphatase activity. This has been tested as follows. Fat-cells were prepared with collagenase using reagents depleted of Ca^{2+} by passage through a column of Chelex 100. The cells were broken and mitochondria separated in Chelex-treated medium (sucrose–Tris–glutathione–albumin containing 2 mM-EGTA). Mitochondria were then incubated and samples extracted and assayed for pyruvate dehydrogenase and glutamate dehydrogenase (as a marker enzyme). The results are shown in Fig. 5. Addition of 2-oxoglutarate and malate (as substrates for respiration)

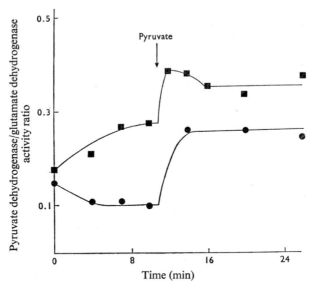

Fig. 5. *Effect of EGTA and CaCl$_2$ on pyruvate dehydrogenase activity in rat epididymal fat-cell mitochondria*

Mitochondria were prepared by the method of Martin & Denton (1970, 1971) but with reagents depleted of calcium by passage through Chelex 100. Mitochondria were incubated at 30°C in air in 125 mM-KCl, 2 mM-potassium phosphate, 20 mM-Tris–HCl, pH 7.4, containing 5 mM-2-oxoglutarate and 0.5 mM-malate and either 2 mM-EGTA (●) or 10 μM-CaCl$_2$ (■). The oxoglutarate and malate were added at zero time, and 1 mM-pyruvate was added as shown after 10 min. Pyruvate dehydrogenase and glutamate dehydrogenase were extracted and assayed as described by Martin *et al.* (1972).

resulted in inactivation of pyruvate dehydrogenase in EGTA medium, presumably as a result of ATP synthesis (Martin et al., 1972). In 10 μM-$CaCl_2$ pyruvate dehydrogenase activity increased slowly. This effect could be due either to activation of the phosphatase by Ca^{2+} or to inhibition of ATP synthesis by Ca^{2+} uptake. Addition of pyruvate led to prompt activation of the dehydrogenase and the response was very similar in the two media. The effect of pyruvate was not therefore dependent upon external Ca^{2+}.

Ruthenium Red and insulin action in fat-pads

Ruthenium Red has been shown to inhibit uptake of calcium by mitochondria from a number of mammalian tissues. In rat liver mitochondria, for example, Ruthenium Red blocks respiration-dependent and ATP-driven uptake of Ca^{2+}, K^+/Ca^{2+} exchange and the binding of Ca^{2+} to high-affinity sites in the mitochondrial membrane (Vasington et al., 1972). We have found with fat-cell mitochondria that O_2 uptake is stimulated by calcium and that this stimulation is dependent on the presence of respiratory substrate, and is enhanced by phosphate and blocked by Ruthenium Red. The incorporation of ^{45}Ca into isolated fat-cell mitochondria was dependent upon respiration and blocked by uncouplers or Ruthenium Red. Uptake of ^{45}Ca could also be supported by ATP in which case it was blocked by oligomycin or Ruthenium Red. In fat-cells, incorporation of ^{45}Ca into the mitochondrial fraction (see following section) was inhibited by Ruthenium Red in the medium (approx. 50% over 10 min; 30% over 30 min and 30% over 60 min). It is not known whether this effect of Ruthenium Red in fat-cells is due to inhibition of fat-cell uptake of ^{45}Ca or whether Ruthenium Red enters fat-cells and inhibits mitochondrial uptake of ^{45}Ca.

The effects of Ruthenium Red on the activation of pyruvate dehydrogenase by insulin and by glucose and fructose in rat epididymal fat-pads has been investigated in experiments with Kirstie de Rivaz (K. de Rivas, R. M. Denton & P. J. Randle, unpublished work). Ruthenium Red did not inhibit activation of the dehydrogenase by glucose or fructose (2 mg/ml). With fructose as substrate, insulin activation of the dehydrogenase was inhibited by Ruthenium Red (5 or 10 μg/ml) in four out of nine experiments. The experimental protocol involved preincubation for 30 min in medium containing fructose (2 mg/ml) with or without Ruthenium Red (5 or 10 μg/ml). Incubations were then made for 30 min in media containing fructose (2 mg/ml) with or without Ruthenium Red in the absence or presence of insulin (10^{-3} i.u./ml). When the results of all experiments were combined (30 observations) the ratios of activities of pyruvate dehydrogenase were: insulin/control 1.64 ± 0.03 (significant activation); insulin+Ruthenium Red/Ruthenium Red alone 1.28 ± 0.03 (effect of insulin); Ruthenium Red/control 1.12 ± 0.03 (small activation by Ruthenium Red); Ruthenium Red + insulin/insulin alone 0.86 ± 0.01 (inhibition by Ruthenium Red in the presence of insulin). The antilipolytic effect of insulin was not affected by Ruthenium Red in the same experiments. These findings could suggest that activation of pyruvate dehydrogenase by insulin with fructose as substrate may involve mitochondrial uptake of calcium and activation of the phosphatase. However, we have no explanation at this time for the somewhat variable inhibition by Ruthenium Red with fructose as substrate. With glucose as substrate Ruthenium

Red did not inhibit insulin action. However, insulin has marked effects on glucose uptake and may increase cell pyruvate with this substrate, whereas the hormone has little effect on fructose uptake and may lower the cell pyruvate concentration with this substrate (Coore et al., 1971). With glucose as substrate, inhibition of pyruvate dehydrogenase kinase by pyruvate may be the dominant factor in activation of the dehydrogenase by insulin.

Exchangeable Calcium in Mitochondria in Fat Cells

These measurements were prompted by the possibility that insulin could activate pyruvate dehydrogenase phosphate phosphatase by increasing mitochondrial calcium content. Such an effect might be detected through an increase in exchangeable calcium in mitochondria and methods have been explored for measurements in fat cells. Fat-cells are advantageous in this respect because they can readily be separated from incubation media by centrifuging through dinonyl phthalate (Gliemann et al., 1972) and mitochondria can be quickly isolated from cells by the method of Martin & Denton (1970, 1971). (The essential features of the procedure are shown in Scheme 1.) Exchangeable calcium has been measured with ^{45}Ca with use of ^{3}H-labelled inulin as an extracellular marker and of glutamate dehydrogenase activity as an index of mitochondrial recovery. Ruthenium Red was used to inhibit further exchange of mitochondrial calcium after breaking the fat-cells and EGTA was used to sequester extramitochondrial calcium.

A number of control experiments have been performed as follows. Exchange is rapid and approx. 80% of the ^{45}Ca incorporation into the mitochondrial fraction in fat-cells takes place in the first 10min. Non-radioactive chase experiments showed that approx. 70% of ^{45}Ca in mitochondria can be removed by 10min of incubation of fat cells in media containing non-radioactive calcium. The efficacy of Ruthenium Red and EGTA in preventing further uptake of ^{45}Ca after cell breakage was shown by adding ^{45}Ca after breaking the cells. These experiments showed that Ruthenium Red and EGTA limit uptake of ^{45}Ca after cell breakage to less than 1% of that incorporated into mitochondria during incubation of cells. In the absence of Ruthenium Red and EGTA, mitochondrial incorporation after breakage was approx. 65%. Ficoll-gradient centrifugation indicated that approx. 90% of the ^{45}Ca is associated with predominantly mitochondrial fractions and

Scheme 1. *Procedure for measurement of exchangeable calcium in mitochondria in fat-cells*

1. (*a*) Isolated rat epididymal fat cells were preincubated for 30min in Krebs–Ringer bicarbonate buffer containing fructose (2mg/ml). (*b*) Cells were then incubated for 30min in medium containing fructose (2mg/ml), [^{3}H]inulin and ^{45}Ca. (*c*) A sample of cells was taken for assay of glutamate dehydrogenase activity per g dry wt.

2. (*a*) Incubation medium was separated from cells by layering on to dinonyl phthalate and by centrifuging at 3000*g* for 1min. (*b*) A sample of medium was taken for assay of d.p.m./ml in [^{3}H]inulin and ^{45}Ca.

3. Cells were broken at 0°C in 0.25M-sucrose, 20mM-Tris–HCl, 2mM-EGTA, 7.5mM-glutathione, Ruthenium Red (5µg/ml), bovine plasma albumin (20mg/ml), pH7.4, by vortex mixing in a glass tube for 2×30s.

4. Fat-plug separated by centrifuging at 1000*g* for 1min. Infranatant assayed for [^{3}H]inulin and ^{45}Ca in total cell fraction.

5. Mitochondrial and supernatant fractions separated by centrifuging at 15000*g* for 4min and assayed for ^{45}Ca, [^{3}H]inulin and glutamate dehydrogenase activity.

Table 2. *Exchangeable calcium in mitochondria in fat-cells*

The experimental conditions were as outlined in Scheme 1. The concentration of calcium in the incubation medium was 1.25 mM and the total volume of incubation medium was 10 ml/g of cells. Infranatant = step 4 in Scheme 1. Supernatant and mitochondrial fractions = step 5 in Scheme 1.

Cell fraction	Space (μl/g of cells) for:		Difference (^{45}Ca − [^{3}H]inulin)
	^{45}Ca	[^{3}H]Inulin	
Total cell (infranatant)	28	12	16
Supernatant	15	10	5
Mitochondria	10.8	0.55	10.3

less than 2% is associated with plasma membranes. Further evidence that the sedimentable ^{45}Ca is incorporated into mitochondria is provided by the observation that much of it is discharged by uncouplers.

The results of a typical experiment are shown in Table 2. The total exchangeable calcium in fat-cells incubated in media containing 2 mM-fructose was 20 nmol/g dry wt. of cells. Approximately two-thirds of this calcium was mitochondrial. The total intracellular water is approx. 20 μl/g dry wt. of cells. The efficacy of the di-nonyl phthalate procedure is shown by the fact that the inulin space of the fat cells was only 12 μl/g of cells, whereas the volume of medium used in incubation was 10000 μl/g of cells. The recovery of mitochondria, on the basis of glutamate dehydrogenase activity, was 90%. The method appears promising for studies of the effects of agents that influence fat-cell dehydrogenase activity on mitochondrial exchangeable calcium.

Conclusions and Summary

These studies would appear to show that pig heart pyruvate dehydrogenase phosphate phosphatase requires Mg^{2+} and Ca^{2+} as cofactors. The same may be true for rat adipose-tissue and pig kidney-cortex phosphatases. The K_m for Mg^{2+} is approx. 1 mM. With calcium–EGTA buffers the K_m for Ca^{2+} is approx. 0.8 μM and full activation is seen with approx. 10 μM-Ca^{2+}. In the absence of EGTA much higher concentrations of Ca^{2+} (approx. 500 μM) are necessary for full activation. The reason for this difference has yet to be ascertained. Calcium appears to act by increasing the affinity of the phosphatase for its substrate, pyruvate dehydrogenase phosphate. Purification of the phosphatase has not been achieved in our laboratory because of instability that develops during purification. This suggests that unknown factors are involved in the stability of the enzyme and resolution of this problem is required before firm conclusions can be drawn regarding the significance of this effect of calcium.

Pyruvate, an inhibitor of pyruvate dehydrogenase kinase, can activate pyruvate dehydrogenase in rat epididymal fat-cell mitochondria *in vitro*, even in the absence of extramitochondrial calcium. The activation of pyruvate dehydrogenase by pyruvate is assumed to involve action of the phosphatase. If this is so, then the activity of the phosphatase in these experiments was not dependent on extramitochondrial calcium. We know of no convincing evidence at this time that movement of Ca^{2+} into mitochondria is involved in the activation of pyruvate dehydrogenase by physiological agents. Experiments with Ruthenium Red (an inhibitor of mitochondrial calcium uptake) have suggested that insulin action may involve calcium uptake but they have not been conclusive.

Evidence has been obtained that ^{45}Ca can be used to measure the exchangeable calcium in mitochondria in fat cells. These measurements have shown that the total exchangeable calcium is approx. 20nmol/g of dry cells and that two-thirds of the cell calcium is mitochondrial. The exchangeable calcium in mitochondria may be as high as 4μmol/ml of mitochondrial water. These observations leave unanswered a number of important questions. Is the Ca^{2+} concentration in mitochondria high enough under all physiological conditions to result in maximum activation of the phosphatase? If so Ca^{2+} may be a cofactor for the phosphatase and not a regulator, and control of the phosphorylation–dephosphorylation cycle may result from modulation of kinase activity. If this is so then recycling may occur under conditions where the kinase is active. As an alternative, is much of the mitochondrial calcium bound by sites of high affinity so that the Ca^{2+} concentration is low and buffered? If this should be the case then mitochondrial Ca^{2+} and phosphatase activity could be modulated by alterations in calcium fluxes across the mitochondrial membrane and/or by changes in affinity of calcium-binding sites. Measurements of exchangeable calcium may be capable of detecting alterations in fluxes and perhaps the use of ionophores in conjunction with EGTA may permit measurements of affinity.

References

Barrera, C. R., Namihara, G., Hamilton, L., Munk, P., Eley, M. H., Linn, T. C. & Reed, L. J. (1972) *Arch. Biochem. Biophys.* **148**, 343–368
Coore, H. G., Denton, R. M., Martin, B. R. & Randle, P. J. (1971) *Biochem. J.* **125**, 115–127
Denton, R. M., Randle, P. J. & Martin, B. R. (1972) *Biochem. J.* **128**, 161–163
Garland, P. B. & Randle, P. J. (1964) *Biochem. J.* **91**, 6c–7c
Gliemann, J., Østerlind, K., Vinten, J. & Gammeltoft, S. (1972) *Biochim. Biophys. Acta* **286**, 1–9
Hucho, F., Randall, D. D., Roche, T. E., Burgett, M. W., Pelley, J. W. & Reed, L. J. (1972) *Arch. Biochem. Biophys.* **151**, 328–340
Jungas, R. L. & Taylor, S. I. (1972) in *Insulin Action* (Fritz, J. B., ed.), pp. 369–413, Academic Press, New York
Linn, T. C., Pettit, F. H. & Reed, L. J. (1969a) *Proc. Nat. Acad. Sci. U.S.* **62**, 234–241
Linn, T. C., Pettit, F. H., Hucho, F. & Reed, L. J. (1969b) *Proc. Nat. Acad. Sci. U.S.* **64**, 227–234
Linn, T. C., Pelley, J. W., Pettit, F. H., Hucho, F., Randall, D. D. & Reed, L. J. (1972) *Arch. Biochem. Biophys.* **148**, 327–342
Martin, B. R. & Denton, R. M. (1970) *Biochem. J.* **117**, 861–877
Martin, B. R. & Denton, R. M. (1971) *Biochem. J.* **125**, 105–113
Martin, B. R., Denton, R. M., Pask, H. T. & Randle, P. J. (1972) *Biochem. J.* **129**, 763–773
Pettit, F. H., Roche, T. E. & Reed, L. J. (1972) *Biochem. Biophys. Res. Commun.* **49**, 563–571
Portenhauser, R. & Wieland, O. (1972) *Eur. J. Biochem.* **31**, 308–314
Randle, P. J. & Denton, R. M. (1973) *Symp. Soc. Exp. Biol.* **27**, 401–428
Roche, T. E. & Reed, L. J. (1972) *Biochem. Biophys. Res. Commun.* **48**, 840–846
Siess, E. A. & Wieland, O. (1972) *Eur. J. Biochem.* **26**, 96–105
Vasington, F. D., Gazzotti, P., Tiozzo, R. & Carafoli, E. (1972) *Biochim. Biophys. Acta* **256**, 43–54
Whitehouse, S. & Randle, P. J. (1973) *Biochem. J.* **134**, 651–653
Wieland, O., Funcke, H. V. & Löffler, G. (1971) *FEBS Lett.* **15**, 295–298

Mitochondrial Uptake of Calcium Ions and the Regulation of Cell Function

By ERNESTO CARAFOLI

Institute of General Pathology, University of Modena, 41100 Modena, Italy

Synopsis

The mitochondrial membrane contains at least three classes of Ca^{2+}-binding sites. The high-affinity sites are of particular interest, since they may be involved in the process of energy-linked Ca^{2+} translocation, possibly as Ca^{2+} carriers. The energy-linked process removes Ca^{2+} from the medium very efficiently, and under special circumstances may lead to the semi-irreversible sequestration of conspicuous amounts of Ca^{2+} within the mitochondria. Under most conditions, however, Ca^{2+} is maintained in mitochondria in a dynamic state, and may be rapidly ejected under the influence of a variety of agents. The affinity and rate characteristics of the energy-linked uptake, together with the fact that Ca^{2+} may be promptly discharged from mitochondria, suggests that mitochondria may act as efficient and important regulators of the concentration of Ca^{2+} in the cytosol. Mitochondria might thus play a role in many Ca^{2+}-sensitive cellular processes, among them the activation and deactivation of contractile systems, the biosynthesis of some complex molecules and the release of several hormones. Through their Ca^{2+}-transporting ability, they might also regulate several membrane-linked functions, as well as key steps of important metabolic pathways.

Introduction

In the 12 years that have elapsed since the fundamental discovery of Vasington & Murphy (1962), mitochondrial Ca^{2+} transport has emerged as one of the most rapidly expanding topics in the field of bioenergetics. Two trends have now become evident. In the first, the properties and the molecular mechanism of the translocation, as well as the components of the Ca^{2+} transfer system, are investigated (Lehninger *et al.*, 1967; Carafoli & Rossi, 1971, 1972). In the second, the transport of Ca^{2+} by mitochondria is considered from the standpoint of its possible role in the regulation of cellular functions (Lehninger, 1970).

The Transport of Ca^{2+} by Mitochondria: Membrane Loading and Matrix Loading

The essential properties of the Ca^{2+} translocation process have been established long ago. The uptake of Ca^{2+} can be energized by the activity of the respiratory chain, or by the hydrolysis of ATP. In the first case, the process is sensitive to inhibitors of the respiratory chain and not to oligomycin, in the second the process is blocked by oligomycin but not by respiration inhibitors. Both systems are

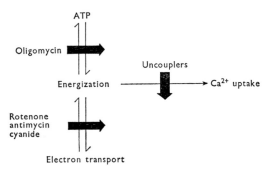

Scheme 1. *Ca^{2+} transport and energy transformations in mitochondria*

inhibited by uncouplers of oxidative phosphorylation, and the scheme shown in Scheme 1 offers the commonly accepted interpretation of the results. Ca^{2+} is thought to discharge the energized state generated by either respiration or ATP, and it is clear from Scheme 1 that Ca^{2+} harvests the energy provided by the respiratory chain before it can be used to synthesize ATP. As will be discussed below, the uptake of Ca^{2+} is thus perhaps the most important alternative to the oxidative phosphorylation of ADP. It has also been found that, under most circumstances, the transport of Ca^{2+} is linked to electron transport by a fixed stoichiometric ratio: about 2.0 calcium ions are taken up as a couple of electrons traverse each one of the three energy-coupling sites along the respiratory chain. Moreover, about 1 H$^+$ leaves the mitochondrion as 1 Ca^{2+} enters. It is important that the energy-linked uptake of Ca^{2+} does not require the simultaneous uptake of permeant anions. Under these conditions, a maximum of about 80–150 nmol of Ca^{2+} are bound to a finite number of anionic sites in the mitochondrial membrane, a process appropriately called 'membrane loading'. When a permeant anion like P$_i$ is also transported into mitochondria, the total amount of Ca^{2+} that can be taken up may exceed 3 μmol/mg of mitochondrial protein, but in this case, most of the Ca^{2+} is accumulated in the matrix as an insoluble phosphate salt ('matrix loading'). The affinity of the energy-linked uptake process for Ca^{2+} is remarkable. Recent measurements by Carafoli & Azzi (1972) have established that the apparent K_m is of the order of 1 μM, that is, a value considerably lower than that for ADP. It is thus easy to understand why isolated mitochondria, presented simultaneously with equivalent amounts of Ca^{2+} and ADP, transport Ca^{2+} first, and phosphorylate ADP only after the largest part of the Ca^{2+} present in the medium has been accumulated (Rossi & Lehninger, 1964). It must also be mentioned here that two specific and extremely effective inhibitors of the energy-linked transport of Ca^{2+} have recently been described. One is La^{3+} (Mela, 1968), the other is a popular histochemical stain, Ruthenium Red (Moore, 1971; Vasington *et al.*, 1972*a,b*), supposedly specific for mucopolysaccharides and glycoproteins. They are maximally active at concentrations of about 1 nmol/mg of mitochondrial protein, which is not far from that of the electron carriers in the respiratory chain.

High- and Low-Affinity Ca^{2+}-Binding Sites

It has been known for many years that some Ca^{2+} is bound by mitochondria even in the complete absence of energy transformations (Rossi *et al.*, 1967;

Wenner & Hackney, 1967). In liver mitochondria, a maximum of about 50nmol of Ca^{2+}/mg of protein are bound in the energy-independent process. The overall affinity is rather low, most measured K_m values being in the 0.1 mM range. Experiments of Scarpa & Azzone (1969) indicate the phospholipids of the mitochondrial membranes as the probable site of the binding, a conclusion supported also by the observation that this type of binding is inhibited by local anaesthetics, which are known to displace Ca^{2+} from phospholipid micelles. La^{3+} and Ruthenium Red also inhibit the binding, but at concentrations which are somewhat higher than those necessary to suppress the energy-linked translocation.

Within the general class of the low-affinity Ca^{2+}-binding sites, Reynafarje & Lehninger (1969) identified a particular category of sites, having very low capacity for Ca^{2+} and very high affinity for it. They have identified these sites from Scatchard plots of Ca^{2+} binding by mitochondria in the absence of energy (Fig. 1), and have found that these sites bind Ca^{2+} with a K_d of 1 μM or less. The important observation has recently been made (Vasington et al., 1972a,b; Lehninger &

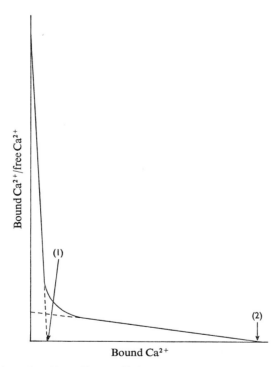

Fig. 1. *High and low-affinity Ca^{2+}-binding sites in rat liver mitochondria*

The experimental conditions are essentially those described by Reynafarje & Lehninger (1969): the reaction medium (0°C, 2ml final volume) contained 0.25 M-sucrose buffered with 10mM-Tris–HCl, pH 7.4, and 5mg of mitochondrial protein. Mitochondria were preincubated with 1 μM-rotenone and 0.25 μg of antimycin A per mg of protein. At 2min after the addition of $^{45}CaCl_2$ (1–80nmol/mg of protein) the suspension was centrifuged at 20000g for 10min, and the distribution of radioactivity between the pellet and the supernatant was measured in a low-background gas-flow counter. Arrows: (1) number of high-affinity sites (0.5–2 sites/mg of protein), (2) number of low-affinity sites (50–70 sites/mg of protein). K_d values for high-affinity sites 0.1–1 μM, those for low-affinity sites 10–100 μM.

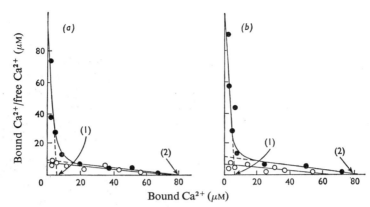

Fig. 2. *Effect of La³⁺ and Ruthenium Red on the high-affinity Ca²⁺-binding sites in rat liver mitochondria*
Experimental conditions as in Fig. 1. Concentrations: La³⁺, 1 μM; Ruthenium Red, 2.5 μM. Arrows: (1) position of high-affinity sites, (2) low-affinity sites. (a) La³⁺; (b) Ruthenium Red.

Carafoli, 1971) that the high-affinity sites are completely blocked by the same low concentrations of La^{3+} and Ruthenium Red which are active in the energy-linked translocation process (Fig. 2). At these very low concentrations, the inhibitors have only a partial effect against the low-affinity sites, which are however severely inhibited by higher concentrations. Another noteworthy property of the high-affinity sites is their absence from some mitochondrial types. Carafoli & Lehninger

Table 1. *Ca^{2+} transport and Ca^{2+}-binding sites in mitochondria*

Energy-linked transport system
  Requires energy
  Shows saturation kinetics
  Specific for Ca^{2+}, Sr^{2+}, Mn^{2+}
  Has high affinity for Ca^{2+} (K_m, about 1 μM)
  Inhibited competitively by Sr^{2+}
  Inhibited by extremely low concentrations of La^{3+} and Ruthenium Red
  Not inhibited by local anaesthetics
  Absent from yeast and blowfly flight muscle mitochondria

High-affinity sites
  No energy required
  Very small number (0.5–5/mg of protein)
  High affinity for Ca^{2+} (K_d, 1 μM or less)
  Specific for Ca^{2+} (Sr^{2+} and Mn^{2+} are also bound)
  Inhibited by extremely low concentrations of La^{3+} and Ruthenium Red
  Not inhibited by local anaesthetics
  Absent from yeast and blowfly flight muscle mitochondria

Low-affinity sites
  No energy required
  Rather large number (up to 70/mg of protein)
  Low affinity for Ca^{2+} (K_d, 10–100 μM)
  Inhibited by local anaesthetics
  Inhibited (not completely) by La^{3+} and Ruthenium Red
  Present also in yeast mitochondria.

(1971) were unable to demonstrate them in blowfly flight muscle and yeast mitochondria, and it is noteworthy that in these mitochondria the energy-linked translocation of Ca^{2+} is either absent or displays very peculiar properties (Carafoli et al., 1970, 1971a). The most important parameters of the energy-linked translocation process discussed so far, together with other parameters that have not yet been mentioned, are listed in Table 1. It is clear that the high-affinity sites and the process of energy-linked translocation share many properties, among them the specificity for Ca^{2+} (Reynafarje & Lehninger, 1969) and the distribution between the inner and outer membrane (Carafoli & Gazzotti, 1973). It is thus permissible to suggest that the binding of Ca^{2+} to the high-affinity sites is a preliminary and necessary step in the sequence of events that leads to its translocation across the inner mitochondrial membrane. It is, however, clear from Table 1 that some of the properties of the low-affinity sites, and of the process of energy-linked translocation, do not coincide, and it has been shown by Mela (1968) that the rate of the energy-linked translocation of Ca^{2+} is greatly stimulated in the presence of concentrations of butacaine sufficient to inhibit very markedly the low-affinity sites. It would thus seem logical to rule out the low-affinity sites as participants in the process of energy-linked translocation. It is however possible, as suggested by Lehninger (1972) that the low-affinity sites, or some of them, are converted into high-affinity sites upon energization of the mitochondrial membrane.

Ca^{2+} and the Oxidation of α-Glycerophosphate: a Third Class of Ca^{2+}-Binding Sites

In addition to the high- and low-affinity sites, a third class of sites capable of interacting with Ca^{2+} have been described by Carafoli & Sacktor (1972). These sites are involved in the oxidation of α-glycerophosphate by the mitochondrial α-glycerophosphate dehydrogenase, which is known (Klingenberg & Buchholz, 1970; Donnellan et al., 1970) to be located only on the external surface of the inner mitochondrial membrane. The Ca^{2+} requirement of the enzyme, which is particularly active in blowfly flight muscle mitochondria, has been demonstrated by Hansford & Chappell (1967) who found that the dehydrogenase exhibits allosteric kinetics with respect to α-glycerophosphate. Ca^{2+} activates the enzyme by lowering its K_m value for the substrate. Carafoli & Sacktor (1972) have found that Ruthenium Red, added to blowfly flight muscle mitochondria in concentrations 2 to 20 times higher than those which would completely suppress the high-affinity binding and the energy-linked translocation of Ca^{2+}, and would also severely decrease the binding to the low-affinity sites, has no effect on the activation of α-glycerophosphate dehydrogenase by Ca^{2+} (Fig. 3). The Ca^{2+}-binding sites involved in the oxidation of α-glycerophosphate are thus completely insensitive to Ruthenium Red in spite of their superficial location on the inner membrane, and are thus clearly different from both the high- and low-affinity sites.

Ca^{2+} Carrier of the Mitochondrial Membrane

It is clear that the parameters of the energy-linked transport system listed in Table 1 correspond closely to the operational definition of a carrier-mediated process. The existence of a specific Ca^{2+} carrier is specially suggested by the fact that

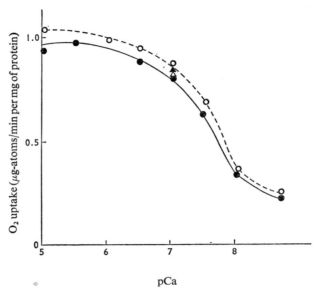

Fig. 3. *The effect of Ruthenium Red on the activation of α-glycerophosphate dehydrogenase by Ca^{2+}*
The preparation of blowfly flight muscle mitochondria and the experimental conditions are described by Carafoli & Sacktor (1972). pCa is $-\log Ca^{2+}$. (○), No addition; Ruthenium Red 0.5nmol (△), 1.6nmol (●), 8.3nmol (▲).

the energy-linked transport system shows saturation kinetics, and by the competitive inhibition of the transport of Ca^{2+} by Sr^{2+} or Mn^{2+}. That the Ca^{2+} carrier is not a member of the respiratory chain, nor chemically and obligatorily linked to electron transport, is strongly suggested by the demonstration that under certain experimental conditions the stoichiometry between Ca^{2+} transport and electron transport may be highly flexible (Carafoli *et al.*, 1966). Also very important is the observation (Selwyn *et al.*, 1970) that Ca^{2+} easily penetrates into mitochondria in the complete absence of respiration if the medium contains high concentrations of Ca^{2+} and permeant anions. The passive penetration of Ca^{2+} down its chemical gradient is inhibited by lanthanides (Selwyn *et al.*, 1970) and by Ruthenium Red (E. Carafoli, unpublished work). As expected of a membrane carrier, the Ca^{2+} carrier of the mitochondrial membrane is capable of transporting Ca^{2+} in either direction down its concentration gradient, but it can also transport Ca^{2+} uphill if a suitable driving force is provided. The bulk of the evidence presently available suggests that the driving force is a respiration-induced electrochemical gradient across the inner mitochondrial membrane (Mitchell, 1968).

Isolation of Ca^{2+}-Binding Proteins (the Postulated Ca^{2+} Carrier) from Mitochondria

The concept of a mitochondrial Ca^{2+} carrier has recently been put to a direct experimental test in a series of investigations carried out by several groups (Lehninger, 1971; Gomez-Puyou *et al.*, 1972; Carafoli *et al.*, 1972*b*; Sottocasa *et al.*, 1972; Carafoli *et al.*, 1973*b*). By using osmotic shock followed by

Table 2. *Characteristics of the Ca^{2+}-binding glycoprotein*

Monomer mol.wt.: about 35000
Sugars: about 5% (sialic acid, hexosamines, neutral sugars)
Phospholipids: variable amounts (up to 1/3 in weight)
Endogenous Ca^{2+}: about 5mol/mol of protein
Endogenous Mg^{2+}: about 3mol/mol of protein
Amino acids: 342 residues (35% glutamic acid and aspartic acid)
Ca^{2+}-binding sites: 2–3/mol (K_d 0.1–1.0 μM) 20–30/mol (K_d 10 μM)
La^{3+} and Ruthenium Red: strong inhibition of Ca^{2+}-binding sites; strong binding of lanthanides

$(NH_4)_2SO_4$ fractionation Gomez-Puyou et al. (1972) extracted from liver mitochondria a glycoprotein which is capable of binding Ca^{2+} with high affinity in a reaction that is sensitive to La^{3+} and Ruthenium Red. The protein contains large amounts of phospholipids and phosphoprotein phosphorous, but is unfortunately made completely insoluble by $(NH_4)_2SO_4$, a fact that renders its further purification and characterization very difficult. In our laboratory we have recently been able to liberate from liver mitochondria, by osmotic shock or by more drastic procedures, a water-soluble glycoprotein which is then purified by preparative electrophoresis on polyacrylamide gels. The protein has been purified and characterized to a very considerable extent (Table 2), and it has been possible to establish that most of it originates from the inner membrane. It contains sialic acid, hexosamines and neutral sugars, has a molecular weight of about 35000, and is very rich in glutamic acid and aspartic acid. As shown in Table 2, it also contains tightly bound Ca^{2+} and a variable amount of phospholipids. It binds Ca^{2+} at two classes of sites, different in capacity and affinity. The K_d of the high-affinity sites is of the order of 1 μM, and it has been calculated that the high-affinity sites of the isolated glycoprotein, which bind 2–3mol of Ca^{2+}/mol, account for the majority of the high-affinity sites of intact mitochondria, if not for all of them. The reaction of Ca^{2+} with the isolated glycoprotein is inhibited very markedly by La^{3+} and by Ruthenium Red, and is insensitive to a variety of other mitochondrial inhibitors (Table 3). Experiments are now under way on the relation of the Ca^{2+}-binding glycoprotein to the process of Ca^{2+} translocation in intact mitochondria. Recent results on mitochondria that are unable to transport Ca^{2+} (blowfly flight muscle, yeast) are particularly encouraging (Carafoli et al., 1973b).

Table 3. *The effect of certain agents on the binding of Ca^{2+} by the mitochondrial glycoprotein*
The glycoprotein was isolated from ox liver mitochondria as indicated by Carafoli et al. (1972b), and Sottocasa et al. (1972). The binding of ^{45}Ca was measured by using the Chelex method, as described by Carafoli et al. (1972b), in the presence of 220nmol of ^{45}CaCl$_2$/mg of isolated glycoprotein. Concentration of La^{3+}, 10 μM, and of Ruthenium Red, 25 μM.

Addition	Ca^{2+} bound (nmol/mg of protein)	% inhibition
None	85	—
La^{3+}	0	100
Ruthenium Red	23	72
Uncouplers		
Oligomycin	no inhibition	
Respiratory inhibitors		
Ionophorous antibiotics		

Some Special Properties of Mitochondrial Ca^{2+} Transport

The most important parameters of the mitochondrial system for the transport of Ca^{2+} have been discussed above. Some other properties of the system which are of particular interest to the problem of its biological role, must now be described. One important property has been mentioned already, namely, the fact that the uptake of Ca^{2+} takes precedence over the phosphorylation of ADP, at least in systems *in vitro* (Rossi & Lehninger, 1964). Also very important is the observation of the universal distribution of the Ca^{2+}-transport system among vertebrate mitochondria (Carafoli & Lehninger, 1971), a fact which established the process as one of their most invariant properties, and which can be contrasted, e.g., with the absence of some of the carriers for the respiratory substrates from some mitochondria. The energy-linked uptake of 'limited' amounts of Ca^{2+}, up to the concentrations that saturate the membrane binding sites (80–100nmol/mg of protein), is not accompanied by functional damage to the mitochondria, which maintain normal respiratory control and phosphorylative efficiency (Rossi & Lehninger, 1964). That energy-linked movements of Ca^{2+} occur in the mitochondria of several tissues *in vivo* has been established by Carafoli and his associates in a series of studies in which radioactive Ca^{2+} was administered to animals whose mitochondria had been previously uncoupled in various tissues *in vivo* (Carafoli, 1967; Patriarca & Carafoli, 1968, 1969; Carafoli & Tiozzo, 1970). They have found that, in all tissues examined, the largest part of the cell Ca^{2+} is concentrated in mitochondria. They have also found that the Ca^{2+} pool in mitochondria is not inert: a large fraction of it is maintained in a very dynamic state, presumably available for the metabolic needs of the cytosol. Similar conclusions on the prominence of the mitochondrial Ca^{2+} pool with respect to the other pools of Ca^{2+} in the cell have been drawn by Borle (1972). As expected, part of the move-

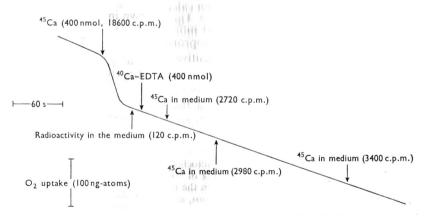

Fig. 4. *Exchange of Ca^{2+} accumulated by rat liver mitochondria with Ca^{2+} of the extra mitochondrial medium*

The O_2 consumption was measured with a Clark polarographic electrode, at 25°C, in a final volume of 3.6ml. The reaction medium contained 80mM-NaCl, 10mM-Tris–HCl, pH7.4, 10mM-sodium succinate and 5mg of mitochondrial protein. The distribution of Ca^{2+} between mitochondria and medium was measured on samples (0.2ml) withdrawn from the polarographic vessel at the times indicated by the arrows and centrifuged at 20000*g* for 10min. The radioactivity was measured in a low-background gas-flow counter.

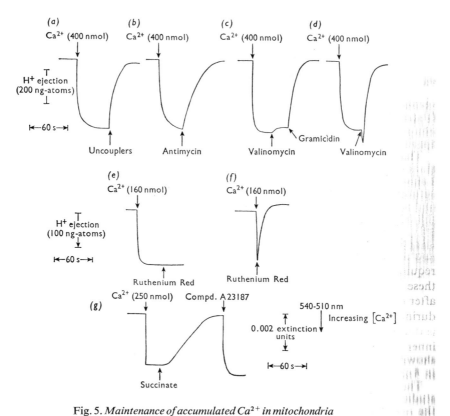

Fig. 5. *Maintenance of accumulated Ca^{2+} in mitochondria*

Effect of uncouplers, respiratory inhibitors, ionophorous antibiotics and Ruthenium Red. In experiments (a)–(f) the movements of Ca^{2+} were followed indirectly by measuring the opposite movements of protons in and out of mitochondria with an expanded scale pH meter. In experiment (g), the movements of Ca^{2+} were followed by monitoring the absorbance changes of the Ca^{2+} indicator murexide in a dual wavelength spectrophotometer (Perkin–Elmer model 356), at 540–510nm. The reaction medium for experiments (a), (b), (c), (e) and (f) contained 80mM-NaCl, 2mM-Tris–HCl, pH7.4, 5mM-sodium succinate and 5mg of mitochondrial protein (rat liver) in a final volume of 1.8ml, at 25°C. In experiment (d), NaCl and sodium succinate were replaced by KCl and potassium succinate. The K^+-specific ionophore valinomycin induced release only in a K^+-containing medium, whereas the ionophore gramicidin was equally active in media containing either Na^+ or K^+. In experiment (g), the reaction medium contained, in a final volume of 2.7ml at 25°C, 0.210M-mannitol, 70mM-sucrose, 10mM-Tris–HCl, pH7.4, 14μM-murexide, 1μM-rotenone and 2.5mg of rat heart mitochondrial protein. Heart mitochondria were prepared in mannitol (0.210M) sucrose (70mM), Tris–HCl, pH7.4 (10mM), EDTA (sodium salt) (0.1mM). The hearts were briefly exposed to about 4mg of Nagarse/g of tissue to improve the yield and the quality of isolated mitochondria. Mitochondria were washed twice, the second time without EDTA. As the traces show, the addition of Ca^{2+} in the absence of respiratory substrates induced a shift in the absorbance of murexide. Initiation of Ca^{2+} uptake by the addition of 10mM-sodium succinate reversed the shift indicating uptake of Ca^{2+}. At this point the addition of the Ca^{2+} ionophore A23187 (kindly donated by Dr. H. Lardy, University of Wisconsin, Wis., U.S.A.) caused a prompt and complete release of Ca^{2+}, shifting again the murexide trace to its original value. Concentrations of inhibitors: uncoupler (carbonyl cyanide *p*-trifluoromethoxyphenylhydrazone), 1μM; antimycin A, 0.25μg/mg of protein; valinomycin and gramicidin, 1μg/mg of protein; compd. A23187, 2μg/mg of protein.

ments of Ca^{2+} which take place in mitochondria *in vivo* are not energy-dependent (Patriarca & Carafoli, 1968). Whether they involve the high-affinity or the low-affinity Ca^{2+}-binding sites of the mitochondrial membrane, or both, cannot be specified at the moment.

That accumulated Ca^{2+} is maintained in mitochondria in a dynamic steady state is shown also by the experiment reported in Fig. 4 (E. Carafoli & A. L. Lehninger, unpublished work). Mitochondria were allowed to accumulate a saturating pulse of radioactive Ca^{2+}. When the accumulation was completed, and no more than 0.1 % of the total Ca^{2+} added was left in the medium, the extramitochondrial pool of Ca^{2+} was expanded by the addition of a second pulse of Ca^{2+}, non-radioactive and equal in magnitude to the first, but complexed with EDTA in a 1:1 ratio. Under these conditions, only marginal amounts of EDTA were bound by mitochondria, no significant net uptake of 'extra' Ca^{2+} occurred, and the free Ca^{2+} concentration in the medium was kept well below the amount required to induce structural damage to mitochondria. Fig. 4 shows that under these conditions the radioactivity in the extramitochondrial medium increased after the addition of the Ca–EDTA pulse (appropriate controls have shown that during the same time the total amount of Ca^{2+} in mitochondria did not vary). It is thus clear that Ca^{2+} taken up by mitochondria, and presumably bound to the inner membrane, exchanges easily with Ca^{2+} of the medium; in the experiment shown in Fig. 4 about one-third of the mitochondrial Ca^{2+} underwent exchange in 5 min.

The maintenance of Ca^{2+} in mitochondria after energy-linked uptake has been studied by Drahota *et al.* (1965). They have established that Ca^{2+} is maintained in the organelles at the expense of the modest degree of energy-coupling which occurs during the phase of resting respiration that follows the activated respiration during the uphill influx of Ca^{2+}. If the energy flow is interrupted, e.g., by uncouplers, the accumulated Ca^{2+} leaves mitochondria down its concentration gradient. As shown in Fig. 5, the uncoupler-induced ejection of Ca^{2+} is complete, and at least as rapid as the uptake, after 'membrane loading'. As expected, however, it is much slower when Ca^{2+} is accumulated in the matrix together with P_i. In addition to uncouplers, a number of agents capable of inducing rapid release of Ca^{2+} from mitochondria have been described, among them respiratory inhibitors (Drahota *et al.*, 1965), various ionophorous antibiotics including the newly discovered Ca^{2+} ionophore A23187 (Carafoli *et al.*, 1969; Reed, 1972), and under some circumstances, Ruthenium Red (Rossi *et al.*, 1973). It is noteworthy that Ruthenium Red induces release only if added while Ca^{2+} is being transported to its binding sites in the inner membrane. No release is observed if Ruthenium Red is added after the energy-linked accumulation of Ca^{2+} has been completed (Fig. 5). The information provided by the experiments with these release-inducing agents has been very important for the demonstration of the reversibility of the Ca^{2+}-transport process, which is perhaps the most important parameter to be considered when discussing a role of mitochondria in the biological regulation of Ca^{2+}. It is clear, however, that the agents mentioned above have no significance from a physiological point of view, and can only be regarded as interesting models. Experiments in which rapid losses of mitochondrial Ca^{2+} are induced by altering the ionic composition of the medium are therefore of particular interest

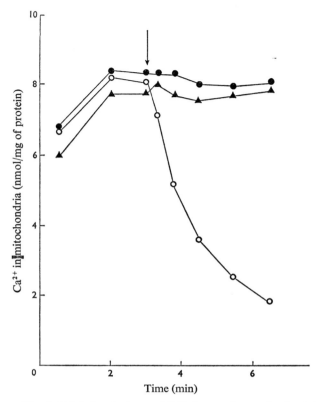

Fig. 6. *Na^+-induced release of Ca^{2+} from rat heart mitochondria*

The reaction medium contained 0.210M-mannitol, 70mM-sucrose, 10mM-Tris–HCl, pH 7, and 5mg of rat heart mitochondrial protein in a final volume of 4ml at 25°C. Mitochondria were preincubated with 1 μM-rotenone and 0.25 μg of antimycin A/mg of protein. $^{45}CaCl_2$ (25nmol) was added to the reaction medium at zero time, and the binding of Ca^{2+} by mitochondria was followed at the times indicated in the figure by Millipore filtration of samples (50μl) of the reaction medium. At the time indicated by the arrow, 50mM-NaCl (○) or KCl (▲) were added. 100mM-mannitol–sucrose–Tris medium was added at the same time to the control (●).

(E. Carafoli, unpublished work). Fig. 6 shows an experiment in which monovalent cations were studied. The releasing effect is apparently specific for Na^+, is fairly rapid, but can only be seen when the amounts of Ca^{2+} previously taken up by mitochondria are very limited (no energy is usually required for the binding of these small amounts of Ca^{2+}).

Mitochondria and the Ca^{2+}-Dependent Activities of the Cell

General considerations

That Ca^{2+} plays a fundamental role in the regulation of many aspects of cell metabolism and activity is a concept that has become firmly established in relatively recent times, and has rapidly become popular. Ca^{2+} is involved in an impressive list of biological phenomena (Table 4) which span from the activation or inhibition of several enzymes located along key metabolic pathways, to the activation and deactivation of several contractile systems, to some important aspects of hormonal regulation and of the biosynthesis of complex molecules, including essential components of biological membranes. Several membrane functions are also known to be influenced by Ca^{2+}, among them those associated with the transmission of the nervous stimulation and with the permeability, communication and contact of cells, as well as those involved in some important functions of the prostaglandins.

For quite some time, the recurring suggestion that mitochondria could regulate one or more of the activities listed in Table 4 was met by skepticism, mainly because early research in the area was centred on the essentially irreversible process of matrix loading. The demonstration that the transport of Ca^{2+} by mitochondria may be rapidly and completely reversible (see above) has obviously made the suggestion very attractive, and has provided a powerful stimulus for research in the area.

The problem at this point is conveniently split into two points. (1) It must be established whether mitochondria could in principle play a role in the activities listed in Table 4, or at least in some of them. (2) Provided that it can be concluded that mitochondria are theoretically qualified to play a role, their actual involvement in one or more of the Ca^{2+}-dependent activities must be demonstrated.

The question of the theoretical adequacy of mitochondria can be answered

Table 4. *Some calcium-dependent activities*

Activation of enzyme systems
    Glycogenolysis (phosphorylase *b* kinase)
    Lipases and phospholipases
    α-Glycerophosphate oxidation
    Pyruvate oxidation
    Succinate oxidation

Inhibition of enzyme systems
    Inhibition of pyruvate kinase
    Inhibition of phospholipid synthesis

Activation of contractile and motile systems
    Muscle myofibryls
    Cilia and flagella
    Microtubules and microfilaments
    Cytoplasmic streaming
    Pseudopod formation

Hormonal regulation
    Formation and/or function of cyclic AMP
    Release of insulin, steroids, vasopressin, catecholamines, thyroxine and progesterone

Membrane-linked functions
    Excitation–secretion coupling at nerve endings
    Excitation–contraction coupling in muscles
    Exocrine secretion (pancreas, salivary glands and HCl in the stomach)
    Aggregation of platelets
    Action potential (nerve and muscle cells)
    $Na^+ + K^+$-activated adenosine triphosphatase of several membranes
    Tight junctions
    Cell contact
    Binding of prostaglandins to membranes

easily in the affirmative, considering the properties of the Ca^{2+}-binding and transport systems which have been described in Table 1. Although the data in Table 1 were obtained on liver or heart mitochondria, they are probably valid for the majority of vertebrate mitochondria. In addition to the parameters listed in Table 1, of particular interest is the total capacity of the energy-linked membrane loading system and of the energy-independent binding sites, which can remove impressive amounts of Ca^{2+} from the environment (see ahead). Some of the processes listed in Table 4, e.g. the contraction of muscle, involve the mobilization of known and rather large amounts of Ca^{2+}, which are, however, well below the capacity of the mitochondrial membrane. The affinity of the mitochondrial energy-linked system and binding sites for Ca^{2+}, particularly that of the high-affinity sites, is also adequate since most of the activities listed in Table 4 are sensitive to Ca^{2+} in the μM concentration range. All considered, then, and although the information available at the moment on the rate and mechanism of the release of Ca^{2+} from the membrane sites is still scarce, no theoretical objections can be raised against the participation of mitochondria in the activities listed in Table 4.

The actual involvement of the mitochondrial Ca^{2+} pump and/or binding sites in the regulation of the Ca^{2+}-dependent activities is far more difficult to prove, and has not yet been demonstrated with absolute certainty in any of the cases mentioned above. (The difficulty is general, and applies to all other subcellular membrane systems.) Some activities have, however, been studied in particular detail, and the evidence obtained, even if it does not conclusively prove the participation of mitochondria, provides extremely convincing arguments in its favour.

Mitochondria in the Contraction and Relaxation of Heart

As is well known, the contraction of muscle requires the binding of Ca^{2+} by troponin, the Ca^{2+} receptor protein in the myofibrils (Ebashi *et al.*, 1968; Fuchs & Briggs, 1968). During the cycle of contraction and relaxation, Ca^{2+} must thus be reversibly moved between troponin and the binding and/or storage sites in the sarcoplasm, and it is clear that the organelle responsible for these reversible movements must be able to lower the external Ca^{2+} concentration below the threshold value of $0.1 \mu M$, above which the myofibrils contract. It must also have a total capacity for Ca^{2+} binding compatible with the total amount of Ca^{2+} exchanged between myofibrils and sarcoplasm during the contraction–relaxation cycles. It must have a very high affinity for Ca^{2+}, necessary to remove Ca^{2+} from troponin, which binds it with a K_d that may be as low as $2 \mu M$ (Ebashi & Endo, 1968). Finally, it must be able to translate the nervous stimulation into a signal for the release of Ca^{2+} into the sarcoplasm. In skeletal muscles, especially fast skeletal muscles, the uptake of Ca^{2+} by sarcoplasmic reticulum satisfies these requirements, and sarcoplasmic reticulum is thus universally indicated as the organelle responsible for the relaxation of the myofibrils. It is also commonly indicated as the organelle that releases Ca^{2+} into the sarcoplasm to initiate contraction, a compelling suggestion in view of the continuity of the triads of the T-system of the Z-line with the longitudinal channels and vesicles of sarcoplasmic reticulum. The suggestion, however, still lacks direct experimental support.

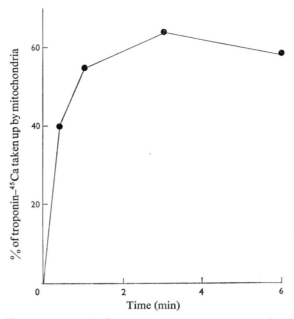

Fig. 7. *Removal of Ca^{2+} from troponin by rat heart mitochondria*

In this experiment (Carafoli, E., Dabrowska, R., Strzelecka, H. & Drabikowski, W., unpublished work) rabbit skeletal muscle troponin was labeled with ^{45}Ca, and the loosely-bound Ca^{2+} was then removed after a brief treatment with Dowex-50. Rat heart mitochondria prepared as in Fig. 6 were incubated at 25°C, with continuous stirring, in a final volume of 2 ml. The medium contained 0.210 M-mannitol, 70 mM-sucrose, 10 mM-Tris–HCl, pH 7.4, 5 mg of mitochondrial protein and 1 mg of ^{45}Ca-labelled troponin. The medium also contained 2.5 mM-Na–ATP and 2.5 mM-$MgCl_2$. Mitochondria were added last to the medium already containing the other components. At the times indicated samples (50 μl) were withdrawn from the medium and filtered through Millipore filters, which arrested mitochondria but not troponin. The filters were then washed with mannitol–sucrose–Tris medium, glued to planchets, and radioactivity was counted in a low-background gas-flow counter.

In heart (and perhaps slow skeletal muscle and smooth muscle, see Batra, 1973) the situation is different. Sarcoplasmic reticulum is scarce, and in addition its isolated elements are less active in Ca^{2+} transport than in fast skeletal muscle (for a review, see Weber, 1966). By contrast, mitochondria are very abundant in heart, and are extremely active in the transport of Ca^{2+}. Their affinity for it (K_d of the energy-linked process, 1 μM, of the high-affinity sites, 0.1–1.0 μM) is clearly adequate for a role in relaxation, and indeed we have shown that heart mitochondria can easily extract Ca^{2+} from troponin (Fig. 7). Their total capacity for Ca^{2+} storage (up to 9 μmol/g of heart in the membrane-loading process, up to 300 nmol/g of heart at the high-affinity sites, up to 4.2 μmol/g of heart at the low-affinity sites) is vastly in excess of the 25 nmol of Ca^{2+} which are exchanged between troponin and sarcoplasm per g of heart per beat. Actually, considering that 1 g of heart corresponds to about 65 mg of mitochondrial protein (Laguens, 1971) mitochondria are required to bind less than 0.5 nmol of Ca^{2+}/mg of protein to account for relaxation. That mitochondria can lower the external Ca^{2+} concentration below 0.1 μM has also been shown long ago by several investigators (Rossi & Lehninger, 1963; Chance, 1965; for a review see Lehninger *et al.*, 1967). In agreement with these results and/or considerations, it might also be added at this point that heart mitochondria are capable of relaxing myofibrils *in vitro*, in a process which is inhibited by Ruthenium Red (Carafoli *et al.*, 1972a). The problem of the release of Ca^{2+} from mitochondria to initiate contraction is certainly more complex, since none of the most widely used release-inducing agents (see above) is physiological. As mentioned already, we have, however, found (E. Carafoli, unpublished work) that under some circumstances Na^+ can become a very powerful release-inducing agent (see Fig. 6). The possibility that Na^+ might represent the messenger that links the incoming nervous stimulation to the mitochondrial membrane system in the interior of the muscle fibre will be of interest.

Mitochondrial Ca^{2+} Transport and Enzyme Regulation

At least three oxidative enzymes in mitochondria are stimulated by Ca^{2+} at very low concentrations: α-glycerophosphate dehydrogenase (Hansford & Chappell, 1967), succinate oxidase (Ogata *et al.*, 1972), and pyruvate dehydrogenase phosphate phosphatase, a key enzyme in the pyruvate dehydrogenase complex (Denton *et al.*, 1972). The latter enzyme is presumably located in the mitochondrial matrix, whereas the other two are firmly bound to the inner membrane, one to its inner, the other to its outer face. Although the information on the species and tissue distribution of the Ca^{2+} sensitivity of these enzyme systems is still not complete, it is clear that the entire operation of the mitochondrial respiratory chain and/or the tricarboxylic acid cycle may be under the control of Ca^{2+}. It is possible that the mitochondrial Ca^{2+}-transport (and binding) system could regulate these enzyme activities by excluding Ca^{2+} from selected zones of the inner membrane, or of the intramitochondrial matrix, or by increasing its concentration in the same areas. The regulation by Ca^{2+} could conceivably be superimposed on that given by ADP, but in some tissues Ca^{2+} might be far more important than ADP as a regulatory agent for the oxidations. It is difficult to see how ADP could become limiting and control respiration in tissues like skeletal muscle or heart, where ATP is continuously dephosphorylated to ADP by creatine phosphokinase.

The role of Ca^{2+} in the control of glycolysis has been especially emphasized by Bygrave (1967). Meli & Bygrave (1972) studied pyruvate kinase, an enzyme which is activated by Mg^{2+} and inhibited by Ca^{2+}, and found in combined experiments with isolated mitochondria and purified pyruvate kinase that the enzyme could be easily 'switched' on and off by inducing or by inhibiting Ca^{2+} uptake by mitochondria. Since liver mitochondria do not accumulate Mg^{2+} (Carafoli *et al.*, 1964), mitochondria do not alter its concentration in the vicinity of pyruvate kinase but alter the Mg^{2+}/Ca^{2+} ratio, thus influencing the activity. By selectively altering the Mg^{2+}/Ca^{2+} ratio in the environment, mitochondria could influence other enzyme systems which are sensitive to this ratio. Among them, the synthesis of phospholipids, which has been studied recently by Porcellati *et al.* (1970a,b). They have found that the incorporation of phosphorylethanolamine into phospholipids was inactivated by Mg^{2+}, and inhibited by Ca^{2+} even in the presence of $12\,\text{mM-}Mg^{2+}$.

Mitochondrial Ca^{2+} Transport and Prostaglandins

As is well known, Ca^{2+} plays a role in some important processes influenced by the prostaglandins, which have been reported to interfere with the permeability of biological membranes to Ca^{2+} (Coceani & Wolfe, 1966; Coceani et al., 1968; Horton, 1969). Kirtland & Baum (1972) found that prostaglandin E_1 increases the rate and extent of Ca^{2+}-induced mitochondrial swelling, and they have therefore suggested that it may act as a mitochondrial Ca^{2+} ionophore. We have found (Carafoli et al., 1973a) that Ca^{2+} greatly stimulates the binding of prostaglandin E_1 to mitochondria (Fig. 8). Prostaglandin E_1 becomes bound during the energy-linked interaction of Ca^{2+} with the mitochondrial membrane, and is released immediately when the active accumulation of Ca^{2+} is completed. It is also noteworthy that anti-inflammatory drugs suppress the Ca^{2+}-induced binding of prostaglandin E_1, probably because they inhibit the energy-linked interaction of Ca^{2+} with mitochondria. Speculations on the physiologial significance of these observations are undoubtedly premature, but there is no reason at the moment to rule out the possibility that mitochondria may be physiological targets for prostaglandin E_1. In particular, the possibility that the inhibition of the binding of prostaglandins to mitochondrial and other cellular membranes may be a component of the anti-inflammatory response seems worth exploring.

Possible Physiological Implications of the Process of 'Matrix Loading'

The assumption has been made so far that the process of matrix loading, in which massive amounts of Ca^{2+} and phosphate are deposited in the mitochondrial matrix, is probably an artificial process, of no interest to cell physiology. Indeed, the process is essentially irreversible (or very slowly reversible), and it is thus very difficult to see how it could be involved in the regulation of metabolism, which requires rapid and reversible changes in the concentration of Ca^{2+}. In addition, the accumulation of large amounts of Ca^{2+} and phosphate, at least by isolated mitochondria, leads to structural and functional damage to the organelles. The situation in the cells, particularly in some cells, may however be different. Several instances have been described in which mitochondria in various cell types accumulate large amounts of Ca^{2+} and phosphate without evident signs of structural damage. Such is, for example, the case of intact toad urinary bladder (Peachey, 1964), or isolated dog papillary muscles (Legato et al., 1968) exposed to high Ca^{2+} concentrations. In both cases, insoluble deposits of Ca^{2+} salts are easily visible within the otherwise normal mitochondrial profiles in the electron microscope. Even more noteworthy are the cases of tissues in whose cells high concentrations of Ca^{2+} are present during normal life, like osteoclasts and osteoblasts, condrocytes in calcifying cartilage, the cells of the calciferous gland of the earthworm or of the egg-shell gland. In the mitochondria of all these tissues, numerous electron-opaque masses, similar to those found in isolated mitochondria after accumulation of Ca^{2+} and phosphate (Greenawalt et al., 1964) are evident in the electron microscope (see Table 5). It is also noteworthy that Cameron et al. (1967) observed an increase in both the number and the size of the intramitochondrial dense granules in the osteoblasts of rats treated with parathyroid hor-

MITOCHONDRIAL UPTAKE OF Ca^{2+}

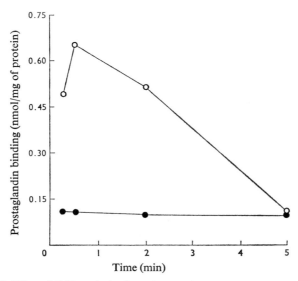

Fig. 8. *Effect of Ca^{2+} on the binding of prostaglandin E_1 to rat liver mitochondria*
The incubation medium contained 100mM-NaCl, 10mM-imidazole buffer, pH 6.4, 10mM-sodium succinate and 5mg of mitochondrial protein in a final volume of 2ml at 25°C. ^{3}H-labelled prostaglandin E_1 (5μM) was added 1min after the addition of mitochondria and Ca^{2+} (250μM) 1min later. Zero time was considered the time at which Ca^{2+} was added. Samples (50μl) were withdrawn from the medium at the times indicated and filtered through Millipore filters, which were then washed and counted for radioactivity in a liquid scintillator. ●, Control without addition of Ca^{2+}; ○, Ca^{2+} (250μM) added.

mone. Thus, there seems to be little doubt that, whenever cells are involved in the process of large-scale deposition or elimination of Ca^{2+}, massive accumulation of Ca^{2+} may occur in their mitochondria. Whether the presence of massive amounts of Ca^{2+} salts in mitochondria is merely the result of increased concentrations of cellular Ca^{2+}, or whether it is a necessary step in the process of bone formation and dissolution, and in the other calcification processes, is at the moment an open question. Lehninger (1970) has suggested that mitochondria could take up soluble Ca^{2+} and phosphate, precipitate them inside the matrix, and then release them again in the form of microcrystalline 'packets' of hydroxyapathite.

Table 5. *Massive uptake of Ca^{2+} by mitochondria in situ*

Condition	Tissue	Reference
Osteoclasts in healing bone fractures	Bone	Gonzales & Karnovsky (1961)
Chondrocytes in calcifying epiphysial cartilage	Cartilage	Martin & Matthews (1969)
Calciferous gland of the earthworm	Epithelium	Crang *et al.* (1968)
Egg-shell gland	Epithelium	Homan & Schraer (1966)
Toad urinary bladder exposed to high Ca^{2+} concentrations	Epithelium and smooth muscle	Peachey (1964)
Dog heart perfused with high Ca^{2+} concentrations	Heart	Legato *et al.* (1968)

Mitochondrial Ca^{2+} Transport and Pathological Calcifications

It is well known that necrotic tissues often undergo calcification. Physicochemical alterations in the necrotic tissue evidently provide a favourable condition for the large scale precipitation of insoluble Ca^{2+} salts. In the necrotic calcification, Ca^{2+} precipitates are found in the interstitial spaces as well as within the dead cells. There is, however, another type of pathological Ca^{2+} deposition, which occurs earlier, well before cell death, and is strictly intracellular. Its possible relevance in the mechanism of cell death was considered by Judah et al. (1964): '... uptake of Ca^{2+} by dying cells is a well known phenomenon, and it is possible that Ca^{2+} uptake is a very early manifestation of cell injury ...'. A very noteworthy characteristic of the early intracellular calcification, as opposed to the necrotic calcification, is its intracellular specificity: Ca^{2+} deposits are found only in mitochondria, the remainder of the cell showing little or no modification of the Ca^{2+} concentration. So far, more than a dozen conditions of cell pathology have been described in which an early and prominent feature is either the presence of electron-opaque masses in mitochondria of various tissues, or the increase of mitochondrial Ca^{2+}. Both spontaneous and experimental pathology are represented (Table 6): coronary occlusion (Herdson et al., 1965; Shen & Jennings, 1972), renal failure (Woodhouse & Burston, 1969), cardiac surgery (D'Agostino & Chiga, 1970), tetanus (Zacks & Sheff, 1964), tumour calcinosis (Lafferty et al., 1965), thioacetamide intoxication (Gallagher et al., 1956), carbon tetrachloride intoxication (Thiers et al., 1960; Carafoli & Tiozzo, 1968), iodoform intoxication (Sell & Reynolds, 1969), uranium intoxication (Carafoli et al., 1971b), isopro-

Table 6. *Massive loading of mitochondria with Ca^{2+} and phosphate in spontaneous and experimental pathology*

	Tissue	Reference
Spontaneous pathology		
Coronary occlusion	Heart	Herdson et al. (1965)
Tumour calcinosis	Tumour	Lafferty et al. (1965)
Renal failure	Heart	Woodhouse & Burston (1969)
Tetanus	Muscle	Zacks & Sheff (1964)
Cardiac surgery	Heart	D'Agostino & Chiga (1970)
Intoxications		
Thioacetamide	Liver	Gallagher et al. (1956)
Carbon tetrachloride	Liver	Thiers et al. (1960)
		Carafoli & Tiozzo (1968)
Plasmocid	Heart and muscle	D'Agostino (1963)
2-Fluorocortisol and sodium phosphate	Heart	D'Agostino (1964)
Papain	Heart	Ruffolo (1964)
Isoproterenol	Heart	Bloom & Cancilla (1969)
Jodoform	Liver	Sell & Reynolds (1969)
Uranium	Kidney	Carafoli et al. (1971b)
Dietary errors		
Mg^{2+} deficiency	Heart	Heggtveit et al. (1964)
Hormonal and vitamin inbalances		
Vitamin E deficiency	Muscle	Van Vleet et al. (1968)
Excess of vitamin D	Kidney	Scarpelli (1965)
Excess of parathyroid hormone	Kidney	Caulfield & Schrag (1964)

terenol intoxication (Bloom & Cancilla, 1969), papain intoxication (Ruffolo, 1964), plasmocid intoxication (D'Agostino, 1963), α-fluorocortisol and sodium phosphate intoxication (D'Agostino, 1964), Mg^{2+} deficiency (Heggtveit et al., 1964), excess intake of parathyroid hormone (Caulfield & Schrag, 1964) or of vitamin D (Scarpelli, 1965). Mitochondrial calcification, even of massive size, is not necessarily followed by cell death, and indeed only a few of the conditions mentioned above are necrotic. In some of the conditions that have been studied most thoroughly, like carbon tetrachloride intoxication in liver (Thiers et al., 1960; Carafoli & Tiozzo, 1968), uranium intoxication in kidney (Carafoli et al., 1971b) and iodoform intoxication in liver (Sell & Reynolds, 1969), the intramitochondrial massive calcification may indeed be reversible. Thus, mitochondrial calcification in cell pathology may be a relatively controlled and transient phenomenon. In other cases, however, it may progress until the concentration of mitochondrial Ca^{2+} becomes incompatible with the functioning of mitochondria. Ca^{2+} in the cytosol then increases, either through mitochondrial lysis, or more simply because mitochondria cease to take up Ca^{2+}, and death of the cell ensues. The accumulation of large amounts of Ca^{2+} by mitochondria in these pathological conditions may represent once again the reaction to an increased concentration of Ca^{2+} in the cytosol, possibly owing to alterations of the permeability of the plasma membrane. It is noteworthy, however, that in the course of carbon tetrachloride and uranium intoxications mitochondria tend to retain their Ca^{2+} more tenaciously than normal mitochondria (Carafoli & Tiozzo, 1968; Carafoli et al., 1971b).

I thank Dr. Roberta Tiozzo, Dr. P. Gazzotti, Dr. C. S. Rossi and Mrs. Daniela Ceccarelli for their collaboration in most of the original experiments described. I am especially indebted to Dr. A. L. Lehninger for many fruitful discussions on the general significance of the process of mitochondrial Ca^{2+} transport. Part of the research described has been supported by the Italian National Research Council (C.N.R.).

References

Batra, A. (1973) *Biochem. Pharmacol.* **22**, 803–809
Bloom, S. & Cancilla, P. A. (1969) *Amer. J. Pathol.* **54**, 373–391
Borle, A. B. (1972) *J. Membrane Biol.* **10**, 45–66
Bygrave, F. L. (1967) *Nature (London)* **214**, 667–671
Cameron, D. A., Paschall, H. A. & Robinson, R. A. (1967) *J. Cell Biol.* **33**, 1–14
Carafoli, E. (1967) *J. Gen. Physiol.* **50**, 1849–1864
Carafoli, E. & Azzi, A. (1972) *Experientia* **28**, 906–908
Carafoli, E. & Gazzotti, P. (1973) *Experientia* **29**, 408–409
Carafoli, E. & Lehninger, A. L. (1971) *Biochem. J.* **122**, 681–690
Carafoli, E. & Rossi, C. S. (1971) *Advan. Cytopharmacol.* **1**, 209–227
Carafoli, E. & Rossi, C. S. (1972) in *Energy Transduction in Respiration and Photosynthesis* (Quagliariello, E., Papa, S. & Rossi, C. S., eds.), pp. 853–868, Adriatica Editrice, Bari
Carafoli, E. & Sacktor, B. (1972) *Biochem. Biophys. Res. Commun.* **49**, 1498–1503
Carafoli, E. & Tiozzo, R. (1968) *Exp. Mol. Pathol.* **9**, 131–140
Carafoli, E. & Tiozzo, R. (1970) *Sperimentale* **120**, 79–92
Carafoli, E., Rossi, C. S. & Lehninger, A. L. (1964) *J. Biol. Chem.* **239**, 3055–3061
Carafoli, E., Gamble, R. L. & Lehninger, A. L. (1966) *J. Biol. Chem.* **241**, 2644–2652
Carafoli, E., Rossi, C. S. & Gazzotti, P. (1969) *Arch. Biochem. Biophys.* **131**, 527–537
Carafoli, E., Balcavage, W. X., Lehninger, A. L. & Mattoon, J. R. (1970) *Biochim. Biophys. Acta* **205**, 18–26
Carafoli, E., Hansford, R. G., Sacktor, B. & Lehninger, A. L. (1971a) *J. Biol. Chem.* **246**, 964–972

Carafoli, E., Tiozzo, R., Ronchetti, I. & Laschi, R. (1971b) *Lab. Invest.* **25**, 516–527
Carafoli, E., Tiozzo, R., Rossi, C. S. & Lugli, G. (1972a) in *Role of Membranes in Secretory Process* (Bolis, L., Keynes, R., & Wilbrandt, W., eds.), pp. 175–181, North Holland, Amsterdam
Carafoli, E., Gazzotti, P., Vasington, F. D., Sottocasa, G. L., Sandri, G., Panfili, E. & de Bernard B. (1972b) in *Biochemistry and Biophysics of Mitochondrial Membranes* (Azzone, G. F., Carafoli, E., Lehninger, A. L., Quagliariello, E. & Siliprandi, N., eds.), pp. 623–640, Academic Press, New York
Carafoli, E., Crovetti, F. & Ceccarelli, D. (1973a) *Arch. Biochem. Biophys.* **154**, 40–46
Carafoli, E., Gazzotti, P., Saltini, C., Rossi, C. S., Sottocasa, G. L., Sandri, G., Panfili, E. & De Bernard, B. (1973b) in *Mechanism in Bioenergetics* (Azzone, G. F., Ernster, L., Papa, S., Quagliariello, E. & Siliprandi, N., eds.), pp. 293–307, Academic Press, New York
Caulfield, J. B. & Schrag, B. A. (1964) *Amer. J. Pathol.* **44**, 365–380
Chance, B. (1965) *J. Biol. Chem.* **240**, 2729–2748
Coceani, F. & Wolfe, L. S. (1966) *Can. J. Physiol. Pharmacol.* **44**, 933–950
Coceani, F., Pace-Asciak, C. & Wolfe, L. S. (1968) in *Worcester Symposium on Prostaglandins*, pp. 39–45, J. Wiley, New York
Crang, R. E., Holsen, R. C. & Hitt, J. B. (1968) *Amer. Inst. Biol. Sci. Bull.* **18**, 299–301
D'Agostino, A. N. (1963) *Lab. Invest.* **12**, 1060–1071
D'Agostino, A. N. (1964) *Amer. J. Pathol.* **45**, 633–644
D'Agostino, A. N. & Chiga, M. (1970) *Amer. J. Clin. Pathol.* **53**, 820–824
Denton, R. M., Randle, P. J. & Martin, B. R. (1972) *Biochem. J.* **128**, 161–163
Donnellan, J. F., Barker, M. D., Wood, J. & Beechey, R. B. (1970) *Biochem. J.* **120**, 467–478
Drahota, Z., Carafoli, E., Rossi, C. S., Gamble, R. L. & Lehninger, A. L. (1965) *J. Biol. Chem.* **240**, 2712–2720
Ebashi, S. & Endo, M. (1968) *Progr. Biophys. Mol. Biol.* **18**, 123–183
Ebashi, S., Kodama, A. & Ebashi, F. (1968) *J. Biochem. (Tokyo)* **64**, 465–477
Fuchs, F. & Briggs, F. N. (1968) *J. Gen. Physiol.* **51**, 655–676
Gallagher, C. H., Gupta, D. N., Judah, J. D. & Rees, K. R. (1956) *J. Pathol. Bacteriol.* **72**, 193–201
Gomez-Puyou, A., Tuena, M., Becker, G. & Lehninger, A. L. (1972) *Biochem. Biophys. Res. Commun.* **47**, 814–819
Gonzales, F. & Karnovsky, M. J. (1961) *J. Cell Biol.* **9**, 299–316
Greenawalt, J. W., Rossi, C. S. & Lehninger, A. L. (1964) *J. Cell Biol.* **23**, 21–38
Hansford, R. G. & Chappell, J. B. (1967) *Biochem. Biophys. Res. Commun.* **27**, 686–692
Heggtveit, A. H., Herman, L. & Mishra, R. R. (1964) *Amer. J. Pathol.* **45**, 757–782
Herdson, P. B., Sommers, H. M. & Jennings, R. B. (1965) *Amer. J. Pathol.* **46**, 367–386
Homan, W. & Schraer, H. (1966) *J. Cell Biol.* **30**, 317–331
Horton, E. W. (1969) *Physiol. Rev.* **49**, 112–161
Judah, J. D., Ahmed, K. & McLean, A. E. M. (1964) *Ciba Found. Symp. Cell. Inj. 1963*, pp. 187–205, Churchill, London
Kirtland, S. J. & Baum, H. (1972) *Nature (London)* **236**, 47–49
Klingenberg, M. & Buchholz, M. (1970) *Eur. J. Biochem.* **13**, 247–252
Lafferty, F. W., Reynolds, E. S. & Pearson, E. S. (1965) *Amer. J. Med.* **38**, 105–117
Laguens, R. (1971) *J. Cell Biol.* **48**, 673–676
Legato, M. J., Spiro, D. & Langer, G. A. (1968) *J. Cell Biol.* **37**, 1–12
Lehninger, A. L. (1970) *Biochem. J.* **119**, 129–138
Lehninger, A. L. (1971) *Biochem. Biophys. Res. Commun.* **42**, 312–318
Lehninger, A. L. (1972) in *Molecular Basis of Electron Transport* (Schulz, J., ed.), pp. 118–130, North-Holland, Amsterdam
Lehninger, A. L. & Carafoli, E. (1971) *Arch. Biochem. Biophys.* **146**, 506–515
Lehninger, A. L., Carafoli, E. & Rossi, C. S. (1967) *Advan. Enzymol.* **29**, 259–320
Martin, J. H. & Matthews, J. L. (1969) *Calcif. Tissue Res.* **3**, 184–193
Mela, L. (1968) *Arch. Biochem. Biophys.* **123**, 286–293
Meli, J. & Bygrave, F. L. (1972) *Biochem. J.* **128**, 415–420
Mitchell, P. (1968) *Chemiosmotic Coupling and Energy Transduction*, pp. 1–111, Glynn Research, Bodmin
Moore, C. (1971) *Biochem. Biophys. Res. Commun.* **42**, 298–305
Ogata, E., Kondo, K., Kimura, S. & Yoshitoshy, Y. (1972) *Biochem. Biophys. Res. Commun.* **46**, 640–645
Patriarca, P. & Carafoli, E. (1968) *J. Cell. Physiol.* **72**, 29–38
Patriarca, P. & Carafoli, E. (1969) *Experientia* **25**, 598–599
Peachey, L. D. (1964) *J. Cell Biol.* **20**, 95–111

Porcellati, G., Biason, M. G. & Arienti, G. (1970a) *Lipids* **5**, 725–733
Porcellati, G., Biason, M. G. & Pirotta, M. (1970b) *Lipids* **5**, 734–742
Reed, P. W. (1972) *Fed. Proc. Fed. Amer. Soc. Exp. Biol.* **31**, 432
Reynafarje, B. & Lehninger, A. L. (1969) *J. Biol. Chem.* **244**, 584–593
Rossi, C. S. & Lehninger, A. L. (1963) *Biochem. Z.* **338**, 698–713
Rossi, C. S. & Lehninger, A. L. (1964) *J. Biol. Chem.* **239**, 3971–3980
Rossi, C., Azzi, A. & Azzone, G. F. (1967) *J. Biol. Chem.* **242**, 951–957
Rossi, C. S., Vasington, F. D. & Carafoli, E. (1973) *Biochem. Biophys. Res. Commun.* **50**, 846–852
Ruffolo, P. R. (1964) *Amer. J. Pathol.* **45**, 741–749
Scarpa, A. & Azzone, G. F. (1969) *Biochim. Biophys. Acta* **173**, 78–85
Scarpelli, D. G. (1965) *Lab. Invest.* **14**, 123–141
Sell, D. E. & Reynolds, E. S. (1969) *J. Cell Biol.* **41**, 736–752
Selwyn, M. J., Dawson, A. P. & Dunnett, S. J. (1970) *FEBS Lett.* **10**, 1–5
Shen, A. C. & Jennings, R. B. (1972) *Amer. J. Pathol.* **67**, 417–433
Sottocasa, G. L., Sandri, G., Panfili, E., De Bernard, B., Gazzotti, P., Vasington, F. D. & Carafoli, E. (1972) *Biochem. Biophys. Res. Commun.* **47**, 808–813
Thiers, R. E., Reynolds, E. S. & Vallee, B. L. (1960) *J. Biol. Chem.* **235**, 2130–2133
Van Vleet, J. F., Hall, B. V. & Simon, J. (1968) *Amer. J. Pathol.* **52**, 1067–1079
Vasington, F. D. & Murphy, J. V. (1962) *J. Biol. Chem.* **237**, 2670–2677
Vasington, F. D., Gazzotti, P., Tiozzo, R. & Carafoli, E. (1972a) in *Biochemistry and Biophysics of Mitochondrial Membranes* (Azzone, G. F., Carafoli, E., Lehninger, A. L., Quagliariello, E. & Siliprandi, N., eds.), pp. 215–228, Academic Press, New York
Vasington, F. D., Gazzotti, P., Tiozzo, R. & Carafoli, E. (1972b) *Biochim. Biophys. Acta* **256**, 43–54
Weber, A. (1966) *Curr. Top. Bioenerg.* **1**, 203–254
Wenner, C. E. & Hackney, J. H. (1967) *J. Biol. Chem.* **242**, 5053–5058
Woodhouse, M. A. & Burston, J. (1969) *J. Pathol.* **97**, 733–745
Zacks, S. I. & Sheff, M. F. (1964) *J. Neuropathol. Exp. Neurol.* **23**, 306–323

Preliminary Studies on the Structure of the Calcium-Activated Adenosine Triphosphatase of Sarcoplasmic Reticulum

By N. M. GREEN, P. M. D. HARDWICKE
and D. A. THORLEY-LAWSON

*National Institute for Medical Research, Mill Hill,
London NW7 1AA, U.K.*

Synopsis

Electron micrographs of vesicles reconstituted from the purified adenosine triphosphatase lipoprotein of sarcoplasmic membranes showed the small projections characteristic of the parent membranes. These projections account for about half of the protein of the molecule of adenosine triphosphatase (mol.wt. 115000). Digestion of the reconstituted vesicles with trypsin gave fragments of molecular weight 55000 and 60000, which were released from the membrane only by treatment with sodium dodecyl sulphate. The larger of these was split more slowly to subfragments of molecular weight 24000 and 33000; the latter carried the phosphorylated active centre of the enzyme. Both of the subfragments and their parent were readily iodinated by lactoperoxidase. The 55000-mol.wt. fragment was only lightly iodinated and probably represents part of the molecule that is protected by interaction with lipid. Removal of lipid from the intact protein by gel filtration in 1% deoxycholate gave an inactive protein that remained in solution after removal of the deoxycholate. Electron micrographs showed small aggregates of particles of diameter about 6nm (60Å). It was not possible to reactivate this protein with lipid, and it is suggested that some irreversible change occurs which leads to loss of much of the hydrophobic surface that normally interacts with the membrane lipids.

It has recently been shown (Racker, 1972) that vesicles reconstituted from the ATPase* lipoprotein, isolated from sarcoplasmic reticulum, regained much of the Ca^{2+}-transporting activity of the parent membrane. In order to provide a structural basis for the mechanism of Ca^{2+} transport we have studied the relation between the ATPase molecule and the membrane by following the effects of removal of lipid on its physical and chemical properties.

The lipoprotein is normally held in solution by small amounts of deoxycholate (MacLennan, 1970). When this is removed, membranous vesicles are re-formed and we have observed, contrary to MacLennan *et al.* (1971), that these vesicles retain the small (60Å × 40Å; 6nm × 4nm) projections seen on the surface of

---

* Abbreviation: ATPase, adenosine triphosphatase.

negatively stained sarcoplasmic vesicles (Greaser et al., 1969). The size of the projection is consistent with its being about half of the ATPase molecule (molecular weight 115000) and suggests that the remainder is within the lipid bilayer, where it can be seen in freeze-etched preparations (MacLennan et al., 1971).

When these vesicles were digested with trypsin (0.5–5 μg/mg of ATPase) and fractionated on sodium dodecyl sulphate–polyacrylamide gels the ATPase band was lost and two fragments of molecular weights 55000 and 60000 appeared in equal amounts. The larger of these fragments was split more slowly to fragments of molecular weights 24000 and 33000. The enzyme was not inactivated, no protein was released from the membrane at this stage and the appearance of the membrane in the electron microscope was unchanged. Separation of the fragments was possible only after dissociation in sodium dodecyl sulphate. The location of the fragments was determined by iodination with ^{125}I catalysed by lactoperoxidase (Marchalonis, 1969). Dissociation of lipid from the intact ATPase in 1% deoxycholate increased its subsequent iodination 2–3-fold, suggesting that part of the molecule was protected by lipid. In the absence of deoxycholate the label was found mainly on the two smaller fragments, suggesting that they constituted the projecting part of the molecule. Phosphorylation of the ATPase with [γ-^{32}P]ATP led to specific labelling of the '60000' and '33000' fragments, showing that the active centre was located on the projecting part of the molecule.

When the ATPase lipoprotein, containing 33% by weight of lipid, is dissolved in 1% deoxycholate the lipid is removed from the protein and the activity is lost irreversibly [Martonosi (1968) and our observations]. We separated the lipid from the protein on a column of Sephadex G-50 in 1% deoxycholate and then removed the deoxycholate with anion-exchange resin. Surprisingly, the lipid-free protein remained in solution. Gel filtration on Sepharose 4B gave an asymmetric peak covering a range of molecular weight from 2×10^5 to 2×10^6; the maximum was at 5×10^5. In the electron microscope the negatively stained protein appeared as irregular, predominantly linear aggregates of particles of diameter about 6nm (60Å). There was little change in the circular-dichroism spectrum of the protein, and both this and its compact globular shape show that it was not extensively unfolded.

The limited tendency of the protein to aggregate contrasted strongly with the behaviour of the parent lipoprotein and suggested a loss of hydrophobic surface when the lipid was removed. Such a loss would be expected to hinder recombination with lipid and this is consistent with our observation that no more than 5% of the original ATPase activity was regained when reconstitution was attempted under the conditions described by Racker (1972).

Although we have not been able to resolve single molecules of the lipoprotein in the electron microscope we concluded from the appearance of the vesicular aggregates that the molecule was elongated, since the projection was 6nm (60Å) long and this was only half of the molecule. We suggest that removal of the lipid and its replacement by detergent leads to an unstable molecule that rearranges to a more globular form in which the exposed hydrophobic surfaces interact intramolecularly. The transformation is not implausible if one postulates that the '55000' fragment, buried within the membrane, possesses an aqueous channel for the transport of Ca^{2+} and that the polar surfaces forming such a channel move to the outside of the molecule when lipid is replaced by detergent.

These experiments, together with further evidence for a structure of this type, will be published in detail elsewhere (Thorley-Lawson & Green, 1974; Hardwicke & Green, 1974).

We thank Mr. D. Julian and Mr. E. J. Toms for technical assistance and Dr. M. V. Nermut and Dr. N. G. Wrigley for much of the electron microscopy.

References

Greaser, M. L., Cassens, R. G., Hoeckstra, W. G. & Briskey, E. J. (1979) *J. Cell. Physiol.* **74**, 37–50
Hardwicke, P. M. D. & Green, N. M. (1974) *Eur. J. Biochem.* in the press
MacLennan, D. H. (1970) *J. Biol. Chem.* **245**, 4508–4518
MacLennan, D. H., Seeman, P., Iles, G. H. & Yip, C. C. (1971) *J. Biol. Chem.* **246**, 2702–2710
Marchalonis, J. J. (1969) *Biochem. J.* **113**, 299–305
Martonosi, A. (1968) *J. Biol. Chem.* **243**, 71–81
Racker, E. (1972) *J. Biol. Chem.* **247**, 8198–8200
Thorley-Lawson, D. A. & Green, N. M. (1974) *Eur. J. Biochem.* in the press

Biochem. Soc. Symp. (1974) 39, 115–132
Printed in Great Britain

Calcium Ions and the Function of the Contractile Proteins of Muscle

By S. V. PERRY

Department of Biochemistry, University of Birmingham, Birmingham B15 2TT, U.K.

Synopsis

Contractile activity in muscle is controlled by regulation of the Mg^{2+}-stimulated actomyosin ATPase (adenosine triphosphatase) of the actomyosin of the myofibril. In higher animals this involves the regulatory protein system, consisting of the troponin complex and tropomyosin. At Ca^{2+} concentrations of $\leqslant 0.10\,\mu M$ this system prevents, in some way not as yet understood, the interaction of actin with the myosin that is essential for the high rate of hydrolysis of $MgATP^{2-}$ associated with contraction. At $\geqslant 10\,\mu M$-Ca^{2+} the calcium-binding protein of the troponin complex becomes saturated with Ca^{2+} and undergoes conformational change and the regulatory protein system is unable to prevent the interaction of actin with myosin that modifies its enzymic properties; $MgATP^{2-}$ is now split at a high rate. The troponin complex is localized with a periodicity of 38.5 nm (385 Å) along the I filament and tropomyosin may play a role in transmitting the effects of Ca^{2+} to actin monomers not in direct contact with the complex. In certain primitive molluscs the troponin complex is absent and the Ca^{2+} regulation of the Mg^{2+}-stimulated ATPase is mediated directly through one of the light chains of myosin. Ca^{2+} has an indirect effect on the myofibrillar proteins through activation of phosphorylase kinase and 'myosin light-chain kinase', enzymes which phosphorylate the troponin complex and a light chain of myosin respectively. As yet no regulatory function has been demonstrated for these phosphorylation processes.

Introduction

The regulation of the contractile process in muscle exhibits exceptional features in that the response is very fast and involves a large increase in enzymic activity. In the fastest muscles, for example, the ATPase (adenosine triphosphatase) activity of the myofibril probably increases by a factor of 100 or more in the few milliseconds required to develop maximum tension. Likewise the enzymic activity decays at a similar rate when the muscle returns to the resting state. The ability of calcium to function as an activator or trigger for biological processes has been adapted to accommodate the special requirements of the regulation of contraction and results in the compartmentalization and movement of relatively large amounts of calcium about the muscle cell.

In resting muscle the bulk of the calcium is localized within the sarcoplasmic reticulum, presumably bound to calsequestrin and the 54000-dalton component, both of which proteins have affinity constants for calcium in the region of 10^4–10^5

(MacLennan & Wong, 1971; MacLennan et al., 1972). When muscle is stimulated the sarcoplasmic reticulum becomes momentarily permeable to calcium by some mechanism not as yet understood. The calcium bound to the binding proteins associated with the reticulum dissociates off to equilibrate with the sarcoplasm, which in resting muscle has a calcium concentration of about $0.10\,\mu\text{M}$. As a result the calcium concentration of the sarcoplasm rises to about $10\,\mu\text{M}$ and saturates the binding sites of the calcium-binding protein of the troponin complex. When this occurs the Mg^{2+}-stimulated ATPase activity of the myofibril rises to a high level and contraction takes place.

In resting muscle the ATP and Mg^{2+} concentrations are in the range of 3–15 mM, i.e. about 10^5 times greater than that of the free calcium concentration, and are little changed by contraction, particularly during short periods of activity. In contrast the calcium concentration may rise to about 100 times the resting value although even at its highest value during contraction it is only about one-thousandth of the Mg^{2+} concentration. Further, the total ionic strength of the system which is about 0.15 is contributed to by a number of other ions amongst which potassium is the major species. The proteins in the sarcoplasm and myofibril, whose biological activity is regulated by the binding of calcium, have therefore evolved highly discriminating capacities for binding Ca^{2+} under conditions in which this bivalent cation concentration does not usually exceed $10\,\mu\text{M}$, despite the fact that Mg^{2+} and K^+ are constantly present in 1 mM and 100 mM concentrations respectively. The proteins of this type that are directly involved in the contraction–relaxation system or which are related in some special way to muscle activity are listed in Table 1.

Excluding the case of phosphorylase kinase where the situation is complicated by its subunit structure (Cohen, 1973; Hayakawa et al., 1973) the data in Table 1 would suggest that on average one high-affinity binding site for Ca^{2+} is associated with 5000–10000 daltons of polypeptide.

As the contractile and regulatory proteins make up to 60% of the total protein of the muscle cell they are present in relatively high molar concentrations. For example myosin and the components of the troponin complex are present at a concentration of $100\,\mu\text{M}$ (calculated on the assumption that the proteins are distributed evenly over the whole cell, which they are not). It is more usual, however, for enzymes involved in general metabolism, e.g. phosphorylase kinase to be present in much lower concentrations, e.g. $1-0.10\,\mu\text{M}$. It follows that the flux of calcium required to regulate the contractile system is much larger, probably 100–1000 times greater, than is required for the regulation of enzymes of general metabolic importance in the sarcoplasm. If the affinity constants and number of binding sites per molecule of all the proteins binding calcium are of a similar order, then over 99% of the calcium mobilized during contraction in the mammalian muscle cell will be sequestered by the contractile system. Fish and amphibian muscle is unusual in that the sarcoplasm contains large amounts of a group of unusual low-molecular-weight proteins that can bind calcium, apparently in free solution, and which are not found in other species (Table 1). These proteins, the parvalbumins (Hamoir & Konosu, 1965; Pechere et al., 1971) represent about 5% of the total muscle protein. Their function is obscure for simple calculation indicates that the total capacity for binding Ca^{2+} is 5–10 times greater than the

Table 1. *Proteins of muscle involved in the binding of Ca^{2+}*

	Mol.wt.	Concn. in cell	Mol of Ca^{2+}/mol	Affinity constant
Myofibril				
Troponin (calcium-binding protein)	18.5×10^3	1×10^{-4}	2	High, 10^6–10^7
Myosin (EDTA light chain) (Molluscs)	18×10^3	$\sim 10^{-4}$		Low
Sarcoplasm				
Phosphorylase kinase	1.3×10^6	3×10^{-6}	24	High, 10^6–10^7
Parvalbumin (fish)	11×10^3	7×10^{-4}	2	High, 10^6–10^7
Sarcoplasmic reticulum				
Interior calsequestrin	44×10^3		43	Intermediate, 10^4–10^5
54000-dalton component	54×10^3		~ 20	Intermediate, 10^4–10^5
Exterior acidic proteins	20×10^3–30×10^3		~ 3	High, $\sim 10^6$

troponin complex assuming that the nature and mode of action of the fish troponins are similar to their mammalian counterparts. The role of the parvalbumins in fish may be to act as a calcium buffer although the special role of this system in the fish is at present unknown.

Calcium and the Regulatory Protein System of the Myofibril

Although calcium is essential for the ATPase and contractile activities of the myofibril *in vivo* it is not essential for enzymic and contractile activities in systems consisting of actin and myosin alone (Perry & Grey, 1956a,b; Schaub et al., 1967). At the appropriate ionic strength carefully purified actomyosin will hydrolyse ATP and superprecipitate in the presence of Mg^{2+} when the EGTA [ethanedioxybis(ethylamine)tetra-acetic acid] concentration is such to ensure that the Ca^{2+} concentration is <10nm. In the myofibril the contractile proteins are associated with the regulatory protein system consisting of tropomyosin and the troponin complex (Scheme 1) which confers Ca^{2+} sensitivity upon the hydrolysis of $MgATP^{2-}$ by actomyosin systems. Thus in resting muscle where the Ca^{2+} concentration is 10–100nm the Mg^{2+}-stimulated ATPase of the myofibril is low

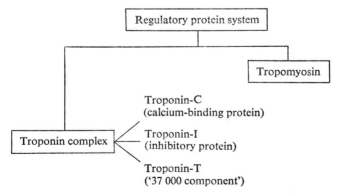

Scheme 1. *Components of regulatory protein system of the myofibril*
For details see the text.

and on stimulation rises to the high rate associated with contraction in response to the increase in Ca^{2+} concentration to about $10\,\mu M$. This type of regulation depends on the special enzymic properties of the actomyosin ATPase. Whereas myosin will hydrolyse ATP at a high rate in the presence of Ca^{2+} at concentrations that ensure that the substrate is $CaATP^{2-}$, under comparable conditions with $MgATP^{2-}$ as substrate the rate of hydrolysis is very low. Actin, however, modifies the enzymic properties of myosin in some way not yet understood, so that $MgATP^{2-}$ is hydrolysed at a high rate. In muscle the ionic relationships are such that virtually all the substrate is $MgATP^{2-}$ and this is therefore only split at a significant rate when the interaction between the actin and myosin occurs. It is this effect of actin on the mechanism of ATP hydrolysis by myosin that is controlled by the regulatory protein system. In the absence of Ca^{2+} the regulatory system prevents actin from interacting with myosin in the manner required for a high rate of hydrolysis of $MgATP^{2-}$. At Ca^{2+} concentrations of $\geqslant 10\,\mu M$ the effect of the regulatory system is lost and the myofibril splits $MgATP^{2-}$ at the high rate required for contractile activity and similar to that obtained with actomyosin in the absence of the regulatory system.

Ebashi (1963) showed that to restore Ca^{2+} sensitivity to the contractile and Mg^{2+}-stimulated ATPase activity of purified actomyosin an additional protein system was required. This was originally considered to be a form of tropomyosin which he named 'native tropomyosin' but it was later shown that the complete regulatory protein system consisted of tropomyosin and a new component, troponin (Ebashi & Kodama, 1965). The latter has now been shown (Hartshorne & Mueller, 1968; Schaub & Perry, 1969; Wilkinson et al., 1971, 1972; Greaser & Gergely, 1971; Ebashi et al., 1971; Ebashi, 1972; Drabikowski et al., 1971; Murray & Kay, 1971) to consist of a complex of three well-defined proteins whose properties and functions as well as those of the other component of the protein regulatory system, tropomyosin, are summarized below.

Calcium-binding protein (troponin-C)

This protein has a molecular weight of 18500 and low isoelectric point. It possesses a high affinity for calcium and its binding sites are probably saturated by two molecules of Ca^{2+} per molecule. It neutralizes inhibition of the Mg^{2+}-stimulated ATPase of actomyosin by the inhibitory protein, with which it can form a complex in the presence of Ca^{2+}.

Inhibitory protein (troponin-I)

The inhibitory protein possesses a molecular weight of 23000 and an isoelectric point in the region of pH 7–8. This protein is a specific inhibitor of the Mg^{2+}-stimulated ATPase of purified actomyosin.

'37000 component' (troponin-T)

The role of this component is less well defined than those of the other two components of troponin. It has a rather similar amino acid composition to the inhibitory protein and a similar isoelectric point. Troponin-T interacts with tropomyosin and is essential for the reconstitution of regulatory protein system to give full restoration of Ca^{2+} sensitivity to the Mg^{2+}-stimulated ATPase of actomyosin.

Tropomyosin

Unlike the components of the troponin complex tropomyosin is a fibrous molecule about 40 nm × 2.5 nm (400Å × 25Å) in dimension. The molecule consists of two polypeptide chains in α-helical conformation which form a coiled coil. It increases the inhibitory activity of the inhibitory protein and is essential for the restoration of calcium sensitivity to the Mg^{2+}-stimulated ATPase of actomyosin by all combinations of the components of the troponin complex.

It has not been demonstrated with precision that the three components of the troponin complex occur in equimolar amounts *in situ*. Nevertheless in most troponin preparations obtained from white skeletal muscle the individual components are present in roughly equimolar proportions and this is presumed to be the case *in situ*. Although all components are required for restoration of activity some calcium sensitivity can be obtained in the absence of troponin-T (see Fig. 1). Nevertheless to restore activity to a value compatible with the undissociated complex it appears that troponin-T is required. The roles of the components of the regulatory protein system are summarized in Fig. 1 where the effects of various combinations on the Mg^{2+}-stimulated ATPase of actomyosin are schematically represented by histograms.

Properties and Mode of Action of the Calcium-Binding Protein of the Troponin Complex

As the target protein for calcium in the regulatory protein system the calcium-binding protein has a high affinity for this bivalent cation. The precise nature and

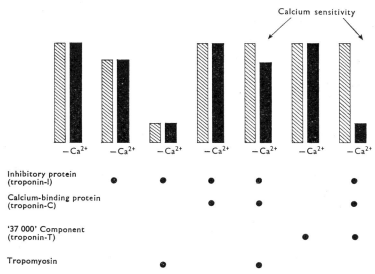

Fig. 1. *Schematic representation of the effects of various combinations of the components of the regulatory protein system on the Mg^{2+}-stimulated ATPase of actomyosin in the presence of Ca^{2+} and EGTA*
ATPase activity of actomyosin measured in 25 mM-Tris–HCl, pH 7.6, 2.5 mM-$MgCl_2$ 2.5 mM-ATP with 0.1 mM-$CaCl_2$ present (hatched histograms), and 2 mM-EGTA present (blocked histograms). The components added to the system are indicated by black dots beneath the pairs of histograms.

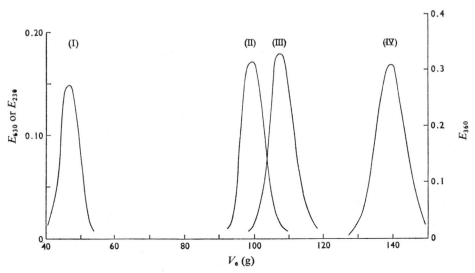

Fig. 2. *Gel filtration of calcium-binding protein in the presence and absence of CaCl$_2$*

Protein (10 mg) in 0.25 ml of 0.1 M-KCl–20 mM-Tris–HCl (pH 7.6)–1 mM-dithiothreitol applied to column of Sephadex G-200 previously equilibrated against the buffer, 20°C. Peaks: (I), Blue Dextran; (II), calcium-binding protein with 4 mM-EGTA included in buffer; (III), calcium-binding protein with 0.1 mM-Ca^{2+} included in buffer; (IV), dinitrophenylglycine. From Head & Perry (1974).

number of sites per molecule of high and medium affinity for Ca^{2+}, i.e. those with affinity constants in the range $10^6 \cdot \text{M}^{-1}$ to $10^7 \cdot \text{M}^{-1}$, is not generally agreed, however. This uncertainty may be due to spontaneous changes in the affinity constants of the Ca^{2+}-binding sites that occur on storage (Perry *et al.*, 1973; M. R. C. Winter, unpublished results) or possibly owing to variations that exist between preparations of the calcium-binding protein for reasons as yet unknown. Hartshorne & Pyun (1971) reported two sites with affinity constants of $5.4 \times 10^6 \cdot \text{M}^{-1}$ and $2 \times 10^5 \cdot \text{M}^{-1}$ respectively. Potter & Gergely (1972) on the other hand report only one calcium-binding site of affinity constant $10^6 \cdot \text{M}^{-1}$ per molecule although they claim that the number increases in the presence of the inhibitory protein. Our studies on spectroscopic changes associated with Ca^{2+} binding are compatible with two sites per molecule although it is uncertain as to whether these sites have similar or markedly different affinities for Ca^{2+} for some variation exists between the preparations (Head & Perry, 1973; Perry *et al.*, 1973).

When calcium is bound a marked conformational change occurs in the calcium-binding protein. This conformational change results in changes in the transport properties, e.g., increased electrophoretic mobility, more retarded elution in gel filtration (Fig. 2), and a change in sedimentation velocity (Murray & Kay, 1972; Head & Perry, 1974). All these changes are compatible with a decrease in the Stokes radius of the molecule. The conformational change is also reflected in changes in the u.v.-absorption spectrum (Fig. 3) and the tyrosine fluorescence (M. R. C. Winter, unpublished results; Perry *et al.*, 1973) that are complete when two molecules of Ca^{2+} are bound per molecule of calcium-binding protein.

Spectral changes are obtained on interaction with a number of different bivalent

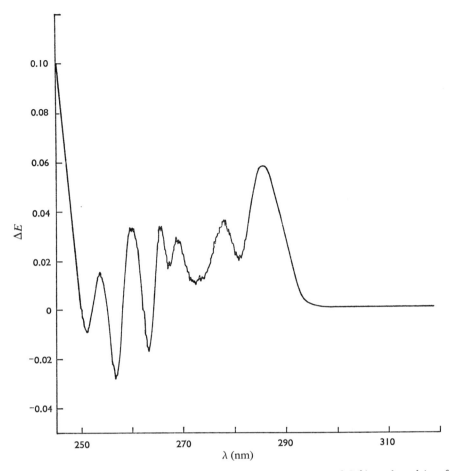

Fig. 3. *Difference spectrum of calcium-binding protein on addition of Ca^{2+} to the calcium-free protein*

In the reference cell (1 cm light-path) calcium-binding protein (6 mg/ml) freed of Ca^{2+} by treatment with Chelex, 100 mM-KCl–10 mM-Tris–7.5 mM-HCl, pH 7.6. The sample cell contained the same constituents with 0.1 mM-CaCl$_2$ added. From Head & Perry (1974).

cations although only with Cd^{2+} are the effects virtually equivalent to those obtained with Ca^{2+}. With other bivalent cations the extent of the changes is not identical with that obtained with Ca^{2+} and much higher amounts of cation are required to bring about maximum change. In the series Ca^{2+}, Cd^{2+}, Sr^{2+}, Mn^{2+}, Mg^{2+}, the spectral changes obtained with Mg^{2+} are least like those obtained with Ca^{2+} (Fig. 4).

In the presence of Ca^{2+} the calcium-binding protein forms a complex with the inhibitory protein that is stable in urea concentrations of 8 M and in the pH range 7.0–8.6. The complex migrates with a characteristic electrophoretic mobility on polyacrylamide gel either in 8M-urea or 40% glycerol at pH 8.0 which is different from that of either of the component proteins (Fig. 5). Gel filtration and cross-linking studies with dimethyl suberimidate and molecular-weight determination by sedimentation-equilibration procedures all indicate that the complex is composed of one molecule of each of the two components (Head & Perry, 1974). On addition of EGTA the complex dissociates into its component proteins. Thus in the presence of Ca^{2+} the calcium-binding protein exists in a form which can interact specifically with the inhibitory protein. From the stability of the complex to ionic strength and general considerations it is unlikely that a calcium bridge is used to link the two proteins together. It is more likely that the conformational change resulting from the binding of calcium exposes or modifies the orientation of amino acid side chains on the calcium-binding protein so that they can interact specifically with groups on the inhibitory protein.

The third feature of the calcium-binding protein that is of functional significance is its property of neutralizing the inhibition of the Mg^{2+}-stimulated ATPase of purified actomyosin by inhibitory protein. Studies with acto-heavy meromyosin indicate that one molecule of calcium-binding protein neutralizes the effect of at

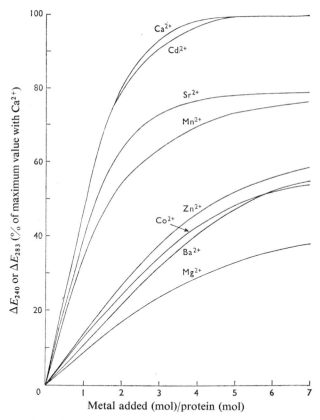

Fig. 4. *Change in absorption at 240nm on addition of divalent cations to metal-free calcium-binding protein*

Calcium-binding protein (1 mg/ml) in 100mm-KCl–10mm-Tris–7.5mm-HCl, pH 7.6. Results plotted at percentage of maximum change obtained when protein is fully saturated with Ca^{2+}. Curves obtained by joining up points obtained by duplicate estimations of absorbance at different bivalent cation concentrations. From Head & Perry (1974).

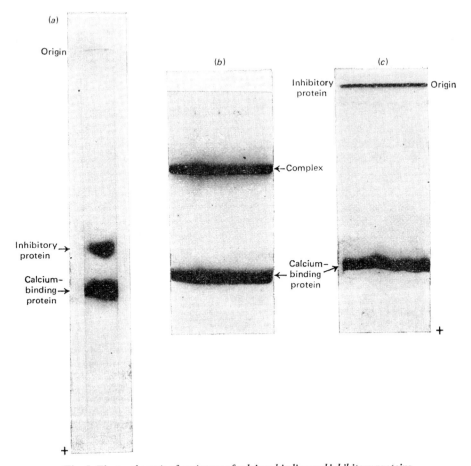

Fig. 5. *Electrophoresis of a mixture of calcium-binding and inhibitory proteins*
(a) Gel electrophoresis of mixture of the inhibitory protein and excess of calcium-binding protein on 10% (w/v) polyacrylamide gels in 100mM-sodium phosphate buffer (pH 7.0) with 0.1% 2-mercaptoethanol and 1% sodium dodecyl sulphate. (b) Mixture of proteins as in (a) electrophoresed in an 8% (w/v) polyacrylamide gel in 6M-urea–25mM-Tris–80mM-glycine (pH 8.6). (c) As for (b) but with 2mM-EGTA added to sample.

least one molecule of inhibitory protein (Perry *et al.*, 1972). This applies whether tropomyosin is present or not. In such systems the neutralization of the action of the inhibitory protein is independent of the presence of Ca^{2+} for it is not affected by EGTA. Somewhat surprisingly this suggests that the formation of the complex between the two proteins that requires the presence of Ca^{2+} is not essential for simple neutralization of the action of the inhibitory protein by the calcium-binding protein.

Although the neutralization of the action of the inhibitory protein by the calcium-binding protein obviously reflects an important aspect of the regulatory function of the troponin complex it is neither Ca^{2+} sensitive nor does it apparently involve complex formation between the components. When the system is fully reconstituted with troponin-T and tropomyosin and exhibits Ca^{2+} sensitivity, complex-formation between inhibitory and calcium-binding proteins presumably plays a part in the Ca^{2+}-regulated neutralization of the inhibitory protein. The properties of the calcium-binding protein which are of significance for its role in the regulatory protein system are summarized in Scheme 2.

Localization of the Calcium-Binding Protein

By studying the localization of ferritin-labelled antibody to the troponin complex with the electron microscope Ohtsuki *et al.* (1967) have concluded that the troponin complex is responsible for the stripes spaced at about 40 nm (400 Å) along the I filament which have been noted by a number of workers (Hall *et al.*, 1946; Draper & Hodge, 1949; Carlsen *et al.*, 1961; Page & Huxley, 1963; Huxley, 1967). More recent electron-microscope studies (Spudich *et al.*, 1972; Hanson, 1973) have confirmed that the troponin complex or at least one or more of the components are localized in this periodic manner. As it is now known that troponin is a complex, the components of which can act as separate antigens (Perry *et al.*, 1972), it cannot be concluded with certainty either from studies with the whole troponin complex or with antibodies to it, which components in fact are localized with the periodicity observed. The conclusions from the investigations with whole troponin, however, are basically correct for by using antibodies to highly purified calcium-binding protein of chicken white skeletal muscle it has been demonstrated that this protein, as is the case with the inhibitory protein, is localized in the I band (Perry *et al.*, 1972; Hirabayashi & Perry, 1973; Ebashi *et al.*, 1972).

The precise localization of the calcium-binding protein has been determined in collaboration with Dr. E. Rome of King's College by X-ray diffraction studies on antibody-labelled muscle fibres. When strips of glycerated rabbit psoas muscle were treated with the antiserum to the calcium-binding protein the only difference in the X-ray diffraction pattern observed compared with that obtained with muscle treated with normal serum was a three- to five-fold increase in intensity of the meridional reflexion corresponding to the repeat of 38.5 nm (385 Å) (Rome *et al.*, 1973). This reflexion is that arising from the actin filament and corresponds very closely to seven times the actin subunit repeat along the axis of the I filament.

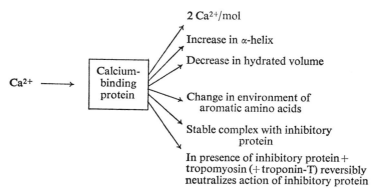

Scheme 2. *Scheme indicating properties of the calcium-binding protein that are calcium-dependent*

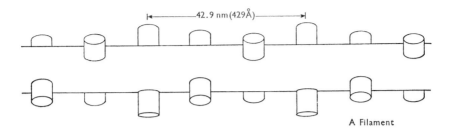
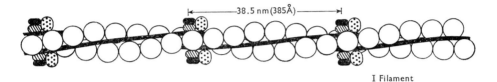

Fig. 6. *Diagrammatic representation of the A and I filaments of the myofibril*
Each cross bridge on the A filament represents the head of the myosin molecule, consisting of two globular subfragment 1 portions of the molecule containing the enzymic and actin-combining centres. The troponin complex is assumed to consist of equimolar amounts of the calcium-binding protein (black), the inhibitory protein (cross-hatched) and troponin-T (stippled). The tropomyosin (dark line) is represented as lying in each groove of the actin filament.

Thus it can be concluded from the X-ray data that the calcium-binding protein is associated with every seventh actin monomer along the I filament. The validity of this conclusion is confirmed by the results of the direct estimation of the amounts of the calcium-binding protein in whole muscle by the isotopic dilution technique. By using ^{14}C-labelled carbamoylmethylated calcium-binding protein from rabbit white skeletal muscle it is estimated that the ratio of the number of molecules of calcium-binding protein to actin monomers, assuming the latter has a molecular weight of 45000 and represents 12% of the total protein of rabbit psoas muscle, is 1:7.3. Somewhat similar values, although determined with less precision, were obtained for red skeletal and cardiac muscle of the rabbit assuming a lower actin content for these tissues (Perry *et al.*, 1973; Head & Perry, 1974).

Thus if these two pieces of data obtained by very different techniques are taken together, it can be concluded that there are two calcium-binding protein molecules arranged every 38.5 nm (385 Å) and as the I filament is composed of a double actin filament, one is probably associated with each of the actin filaments. This arrangement is represented diagrammatically in Fig. 6 where it is assumed that one molecule of inhibitory protein and '37000 component' are present with each molecule of the calcium-binding protein.

As the target protein for Ca^{2+} is localized every 38.5 nm (385 Å) only one, possibly two, actin monomers out of every seven is in either direct or indirect contact through the whole troponin complex, with a molecule of calcium-binding protein. For high ATPase activity each of the actin monomers in the I filament must be potentially able to interact with the active centres of the myosin on the A filament. Thus every actin monomer must be under the control of the periodically localized troponin complex. In the case of the inhibitory protein there is evidence that its inhibitory activity may be transmitted to neighbouring actin monomers through tropomyosin (Perry et al., 1972). It is not unreasonable to presume that the effect of the whole troponin complex is extended by a similar mechanism to actin monomers with which it is not in direct contact. Thus the conformational changes occurring on binding of Ca^{2+} by the calcium-binding protein could produce effects in the inhibitory protein that are transmitted to neighbouring actin monomers in a similar manner to that which tropomyosin extends the action of inhibitory protein when it is effective in the absence of other components of the troponin complex.

At present little is known about the integrated function of the regulatory protein system but clearly the interaction between the components is more subtle than is yet apparent from the studies made on the isolated components. For example, as has been discussed above, the neutralization of the inhibitory protein by addition of the calcium-binding protein is insensitive to Ca^{2+} in contrast with the situation as it exists in the whole complex in the presence of tropomyosin.

The stoicheiometry of the system with respect to Ca^{2+} is of some interest. Full activation of the myofibrillar Mg^{2+}-stimulated ATPase occurs at $10\,\mu M$-Ca^{2+}. At this concentration the calcium-binding protein would be expected from studies on the isolated protein to be fully saturated with two Ca^{2+} bound per molecule, the conformational change to be complete and the complex formed with inhibitory protein. Thus two Ca^{2+} will be bound per functional unit of the I filament consisting of one molecule of troponin complex composed of one molecule of the individual components, seven actin monomers and one tropomyosin molecule. According to the studies *in vitro* of Bremel & Weber (1972) with actin filaments containing amounts of the regulatory protein comparable with those present *in vivo*, four Ca^{2+} ions are bound to the whole system when activation is maximal at $10\,\mu M$-Ca^{2+}. This suggests a large difference between the binding capacity of the isolated calcium-binding protein and that of the whole system. Assuming that the difference cannot be explained by technical errors in measuring calcium binding in the two systems, a possible explanation is that the binding properties of calcium-binding protein are modified in association with other components of the complex. This would necessitate an increase in the number of calcium-binding sites over that existing on the isolated calcium-binding protein. Although this possibility at first sight appears rather unlikely it cannot be ruled out—indeed some preliminary evidence exists that this may be the case (Potter & Gergely, 1972).

Phosphorylation of Troponin

Apart from its intrinsic Ca^{2+}-binding properties the troponin complex is also indirectly affected by changes in the Ca^{2+} concentration of the sarcoplasm. As isolated from white skeletal muscle of the rabbit the troponin complex contains approximately 1 mol of $P_i/80000\,g$ (Perry & Cole, 1973). By prolonged incubation with phosphorylase kinase an additional 2 mol of P_i can be transferred/ $80000\,g$ of troponin. The rate of phosphorylase is much slower than when phosphorylase *b* is the substrate nevertheless phosphorylase kinase is presumably responsible for the phosphorylation of troponin that occurs *in vitro* (see 'Note added in proof'). In whole troponin little phosphorylation of the inhibitory protein occurs for at Ca^{2+} concentrations essential for phosphorylase kinase activity, the calcium-binding protein will be complexed with the inhibitory protein and in the complex the latter is not phosphorylated to any significant extent (Perry & Cole, 1974). Stull et al. (1972) report that the inhibitory protein is phosphorylated to a greater degree than the troponin-T in troponin-B preparations. This occurs because in the absence of calcium-binding protein from troponin-B preparations that contain troponin-T and inhibitory protein, only the latter can be phosphorylated. The extent of phosphorylation of the inhibitory protein is much less in whole troponin, where the troponin-T is the principal substrate for phosphorylase kinase. It is unlikely, however, that the inhibitory protein is normally phosphorylated *in vivo* for the reasons stated above.

The significance of phosphorylation of the troponin complex is not clear particularly in view of the relatively low affinity of phosphorylase kinase for this substrate. There is evidence that enzymes exist in muscle which will dephosphorylate the phosphorylated forms of the inhibitory protein (England et al., 1972) and the '37000 component' but as yet there is no information as to whether a cycle of phosphorylation and dephosphorylation of troponin is associated with contractile activity.

Involvement of Ca^{2+} in Regulatory Systems not Involving Troponin

The troponin system of regulation is widespread in the striated muscles of higher animals although it has been studied mainly on white skeletal muscle, particularly that of the rabbit. In cardiac and red skeletal muscle some differences in the electrophoretic and immunochemical properties of the components and possibly in their molecular weights have been briefly reported (Dabrowska et al., 1973; Head & Perry, 1974; Greaser et al., 1972; Hirabayashi & Perry, 1973). Much less work has been carried out on smooth muscle but there are indications that a similar system is involved. On the other hand components corresponding to those of troponin in the muscle of higher animals cannot be identified in the smooth and striated adductor muscles of certain molluscs, e.g. *Aequipecten irradians* and *Placopecten magellanicus*. Examination of electrophoretograms of extracts of isolated thin I filaments indicates the absence of components corresponding to the troponin-C, troponin-I and troponin-T, in contrast with the appearance of electrophoretograms of similar extracts from mammalian striated muscle (Lehmann et al., 1972). Tropomyosin is, however, present in I filaments from adductor muscle suggesting that this protein is of earlier evolutionary origin and may have a function in the I filament other than as a component of the regulatory system.

Despite the absence of the troponin complex, the Mg^{2+}-stimulated ATPase of adductor muscle actomyosin is Ca^{2+}-sensitive. Thus although the triggering of contractile activity by Ca^{2+} occurs in this tissue it is effective by a mechanism different from that present in the muscles of higher animals. In the case of the adductor muscle the Ca^{2+} sensitivity of actomyosin depends on one of the light-chain components of the myosin present in this tissue (Lehmann et al., 1972). Scallop myosin for example contains about three light chains per molecule of molecular weight approximately 18 000. A light-chain component, of which there is one molecule present per molecule of myosin, is removed by extraction with a dilute EDTA solution. This component, the EDTA light chain, is essential for Ca^{2+} sensitivity of the scallop actomyosin for on its selective removal of the Mg^{2+}-stimulated ATPase of the system it is no longer inhibited by EGTA and behaves essentially like the desensitized actomyosin of mammalian systems. The other light-chain fraction of which there are approximately two molecules per molecule of myosin is removed by thiol reagents with the loss of ATPase activity of the myosin.

The EDTA light chain is not strictly comparable with the calcium-binding protein of the troponin complex in function for when isolated from adductor myosin it does not exhibit special Ca^{2+}-binding characteristics and about two-thirds of the original Ca^{2+}-binding capacity is retained by scallop myosin after its removal. It has been suggested that the Ca^{2+} is bound to the heavy chains of myosin and the EDTA light chain modifies the binding of Ca^{2+} by the myosin in some way, probably by a conformational change (Lehmann et al., 1972).

Despite the differences in the fine mechanism of Ca^{2+} regulation which exist between scallop adductor and the muscles of higher animals the hydrolysis of $MgATP^{2-}$ is regulated in both cases by preventing that interaction of actin with the myosin active centre that is essential for the hydrolysis of $MgATP^{2-}$ at a high rate. In primitive muscles the interaction appears to be broken or inhibited in a manner not understood by the direct effect of Ca^{2+} on the myosin molecule itself. A possible mechanism might involve, for example, the blocking of the active site of myosin by the EDTA light chain in the absence of Ca^{2+}. In the presence of Ca^{2+} conformational change could occur either in the myosin heavy chain which affects the enzymically active site, or on the EDTA light chain itself, which results in normal interaction of actin with the site and leads to a high rate of ATP hydrolysis.

Although such a myosin-regulated system appears to be adequate for the slower, more primitive types of muscle a different system has evolved, perhaps to meet the special requirements for faster response in the muscles of the higher animals. Calcium has been retained as the trigger for activity but the regulatory process has been transferred to the I filament and the interaction of myosin and actin blocked at sites on the actin as described above. One of the consequences of the regulatory system functioning through the actin on the I filaments is that the number of potential sites which need to be regulated are increased. Rather than increase the number of regulatory units to one per actin monomer which would make the I filament very bulky, some co-operative property has been evolved. Tropomyosin which is present in the I filament, and in the more primitive muscles has a purely structural role and has been adapted for this function. In this way every actin monomer becomes under the control of a very much smaller number of regulatory units.

Other Possible Regulatory Roles of Ca^{2+} in Myofibrillar Function

Most of the work so far reported involves Ca^{2+} regulation of the Mg^{2+}-stimulated ATPase of actomyosin with the target protein for calcium being either the troponin system, i.e. I-filament regulation or A-filament regulation on the myosin itself. There is preliminary information, however, that in some species e.g. the annelid worms, that both systems may co-exist in the same muscle (Lehmann et al., 1972). In higher animals particularly the mammals, there is as yet no evidence for a system other than that involving troponin. There is little doubt that the troponin–tropomyosin system is the major mechanism for regulation of ATPase activity of the myofibril but it is sometimes questioned whether this system is adequate enough to decrease the amount of ATPase activity to that which is presumed to exist in resting muscle. Certainly in isolated myofibrils inhibition of the Mg^{2+}-stimulated ATPase by more than 95% is difficult to attain simply by removing Ca^{2+} from the system. In resting muscle it is likely that the rate of ATP hydrolysis is less than about 1% of the maximum rate occurring during contraction. This difference between the enzymic activities obtained *in vitro* and that presumed to exist in *in vivo* may simply be due to the partial disorganization which must occur as a result of isolation and reconstitution of the systems involved. Nevertheless the presence of an additional system of regulation leading to fine control of the contractile response would not be inconsistent with the findings with other biochemical systems, the regulation of which have been studied in detail. The phosphorylation of the troponin components discussed above might be involved in a regulatory system of this type. Such a conclusion is, however, presumptive for there is yet no direct evidence for such a role.

In view of the fact that in primitive muscle types regulation takes place on the myosin itself and involves a light-chain component it is of interest to determine whether the more primitive type of regulatory control is also present in the higher animals that possess the troponin–tropomyosin regulatory system. The role of the light chains of myosin in muscles of the higher animals is as yet unknown but the fact that they vary in number and structure in muscle types containing myosin of different enzymic capacity implies that they may be involved in determining the enzymic properties of the protein. It follows also from the fact that in more primitive muscle systems the light-chain component is apparently able to modify the myosin–actin interaction that such a role might be a more general function of one at least of the light chain fractions which is preserved during evolution.

The recent demonstration (Perrie et al., 1972a,b, 1973a) that enzyme systems exist for the phosphorylation of one of the light-chain components of myosin raises the question whether this process may be involved in some type of regulatory process. There are a number of examples in which phosphorylation controls the activity of an enzyme (Krebs et al., 1966; Linn et al., 1969) but from preliminary studies phosphorylation of the 18 500-molecular-weight light chain of rabbit white skeletal muscle myosin does not appear to have any marked effect on the Ca^{2+}-stimulated ATPase of myosin alone (Perrie & Perry, 1970). If phosphorylation of myosin has a regulatory role it would, however, probably be more apparent in the Mg^{2+}-stimulated ATPase in the presence of actin and possibly the regulatory system, an aspect that has not yet been explored.

The process is highly specific and one of the two serine residues of the sequence Arg-Ala-Ala-Ala-Glu-Gly-Gly-Ser-Ser-Asn-Val-Phe is phosphorylated in each of the two molecules of the 18 500-dalton component of rabbit white skeletal muscle (Perrie et al., 1973a). A similar light-chain component present in the myosin from the white skeletal muscles of other species is likewise phosphorylated.

Phosphorylation also occurs on a light-chain component of red skeletal and cardiac muscle which is of about 19000 molecular weight and presumably slightly different in structure to the 18500-dalton light-chain component of myosin from white skeletal muscle (Perrie *et al.*, 1973b). These facts taken with the recent preliminary report that a light chain present in myosin isolated from platelets also undergoes phosphorylation (Adelstein & Conti, 1973) suggests that a light chain that can be phosphorylated is present in all types of myosin.

Although the original demonstrations of phosphorylation of myosin were made with phosphorylase kinase preparations a number of observations indicated that there were differences between the system phosphorylating myosin and that converting phosphorylase *b* into phosphorylase *a*. It is now clear that the activity obtained was due to an enzyme present in crude preparations of phosphorylase kinase that is not identical with the latter enzyme or with protein kinase (E. M. V. Pires, M. A. W. Thomas & S. V. Perry, unpublished observations, 1973). This enzyme which has not previously been described has been provisionally named 'myosin light-chain kinase' until such time as it is more fully characterized. The fact that it is maximally active at physiological pH values and requires Ca^{2+} for full activity implies that phosphorylation of the 18 500-dalton light chain of myosin can only occur during contraction. These properties together with the ready dephosphorylation that occurs in muscle extracts strongly suggest that the 18000–19000-dalton light chain found in all myosin types examined so far may undergo a cyclical phosphorylation and dephosphorylation during the contractile cycle.

In summary calcium interacts in two main ways with the myofibrillar protein (Scheme 3). The first is by direct interaction with the calcium-binding protein in higher animals and with myosin itself in striated and smooth muscles of certain more primitive organisms. In both examples which represent major regulatory

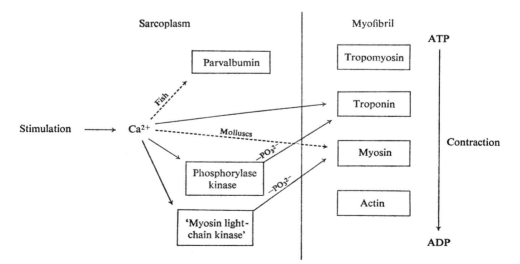

Scheme 3. *Scheme summarizing the relationship between Ca^{2+} and the myofibrillar proteins*

systems in muscle, the Ca^{2+} is bound directly by the myofibril. These systems probably involve more than 99% of the total Ca^{2+} flux between the reticulum and the extra reticular space.

In the second type of interaction a small fraction of Ca^{2+} not bound by the myofibril serves to activate two enzymes in the sarcoplasm namely phosphorylase kinase and 'myosin light-chain kinase'. The former can phosphorylate the troponin-T component of the troponin complex and although troponin-I can be phosphorylated under certain conditions *in vitro* for the reasons stated earlier it is probably not phosphorylated under normal conditions. The 'myosin light-chain kinase' appears to be specific for the myosin component of the myofibril. Although as yet there is no direct evidence for a regulatory function for these two phosphorylations it follows from the properties of the enzymes that phosphorylation can only occur at a significant rate when the Ca^{2+} concentration rises to the value associated with contraction.

Thus in the muscle of the higher animals at least there are changes in three of the proteins of the myofibrils triggered off by a rise in sarcoplasmic Ca^{2+} concentration. The direct effect produces conformational change at the target protein which regulates the Mg^{2+}-stimulated ATPase of actomyosin in a very marked fashion. The other indirect effect can lead to covalent modification of two other protein components of the myofibril. Although evidence of the role of the covalent modification is lacking, analogy with other systems in which cyclical phosphorylation occurs, implies a regulatory function. An attractive hypothesis would be that they represent a mechanism of fine control of the sudden and massive changes produced by the trigger-like response of the Mg^{2+}-stimulated ATPase when the calcium-binding protein of the troponin complex is saturated with Ca^{2+}.

Note Added in Proof

Phosphorylation by another kinase enzyme contaminating the phosphorylase kinase cannot be ruled out at this stage, for phosphorylase kinase preparations will phosphorylate both troponin-T and troponin-I in the presence of EGTA (H. A. Cole & S. V. Perry, unpublished results).

The experimental work described in this article was supported in part by research grants from the Medical Research Council and from the Muscular Dystrophy Associations of America Incorporated.

References

Adelstein, R. S. & Conti, M. A. (1973) *Fed. Proc. Fed. Amer. Soc. Exp. Biol.* **32**, Abstr. 569
Bremel, R. D. & Weber, A. (1972) *Nature (London) New Biol.* **238**, 97–101
Carlsen, F., Knappeis, C. G. & Buchtal, F. (1961) *J. Biophys. Biochem. Cytol.* **11**, 95–117
Cohen, P. (1973) *Eur. J. Biochem.* **34**, 1–14
Dabrowska, R., Dydynska, M., Szpacenko, A. & Drabikowski, W. (1973) *Int. J. Biochem.* **4**, 189–194
Drabikowski, W., Dabrowska, R. & Barylko, B. (1971) *FEBS Lett.* **12**, 148–152
Draper, M. H. & Hodge, A. J. (1949) *Aust. J. Exp. Biol. Med. Sci.* **27**, 465–503
Ebashi, S. (1963) *Nature (London)* **200**, 1010
Ebashi, S. (1972) *J. Biochem. (Tokyo)* **72**, 787–790
Ebashi, S. & Kodama, A. (1965) *J. Biochem. (Tokyo)* **58**, 107–108
Ebashi, S., Wakabayashi, T. & Ebashi, F. (1971) *J. Biochem. (Tokyo)* **69**, 441–445
Ebashi, S., Ohtsuki, I. & Mihashi, K. (1972) *Cold Spring Harbor Symp. Quant. Biol.* **37**, 215–223

England, P. J., Stull, J. T. & Krebs, E. G. (1972) *J. Biol. Chem.* **245**, 6828–6834
Greaser, M. L. & Gergely, J. (1971) *J. Biol. Chem.* **246**, 4226–4233
Greaser, M. L., Yamaguchi, M., Brekke, C., Wotter, J. & Gergely, J. (1972) *Cold Spring Harbor Symp. Quant. Biol.* **37**, 235–249
Hall, C. E., Jakus, M. A. & Schmitt, F. O. (1946) *Biol. Bull.* **90**, 32–50
Hamoir, G. & Konosu, S. (1965) *Biochem. J.* **96**, 85–97
Hanson, J. (1973) *Proc. Roy. Soc. Ser. B* **188**, 39–58
Hartshorne, D. J. & Mueller, H. (1968) *Biochem. Biophys. Res. Commun.* **31**, 647–653
Hartshorne, D. J. & Pyun, H. Y. (1971) *Biochim. Biophys. Acta* **229**, 698–711
Hayakawa, T., Perkins, J. P., Walsh, D. A. & Krebs, E. G. (1973) *Biochemistry* **12**, 567–573
Head, J. F. & Perry, S. V. (1973) *Biochem. Soc. Trans.* **1**, 858–859
Head, J. F. & Perry, S. V. (1974) *Biochem. J.* **137**, 145–154
Hirabayashi, T. & Perry, S. V. (1973) *Abstr. Symp. Calcium-Binding Proteins Warsaw*
Huxley, H. E. (1967) *J. Gen. Physiol.* **50**, *Suppl.* 71–81
Krebs, E. G., Delange, R. J., Kemp, R. G. & Riley, W. D. (1966) *Pharmacol. Rev.* **18**, 163–171
Lehmann, W., Kendrick-Jones, J., Szent-Gyorgyi, A. (1972) *Cold Spring Harbor Symp. Quant. Biol.* **37**, 319–330
Linn, T. C., Pettit, F. H. & Reid, L. J. (1969) *Proc. Nat. Acad. Sci. U.S.* **62**, 234–241
MacLennan, D. H. & Wong, F. T. S. (1971) *Proc. Nat. Acad. Sci. U.S.* **68**, 1231–1235
MacLennan, D. H., Yip, C. C., Iles, G. H. & Seeman, P. (1972) *Cold Spring Harbor Symp. Quant. Biol.* **37**, 469–477
Murray, A. C. & Kay, C. M. (1971) *Biochem. Biophys. Res. Commun.* **44**, 237–244
Murray, A. C. & Kay, C. M. (1972) *Biochemistry* **11**, 2622–2627
Ohtsuki, I., Masaki, T. & Nonomura, Y. & Ebashi, S. (1967) *J. Biochem. (Tokyo)* **66**, 645–650
Page, S. G. & Huxley, H. E. (1963) *J. Cell Biol.* **19**, 369–390
Pechere, J. F., Capony, J. P. & Ryden, L. (1971) *Eur. J. Biochem.* **23**, 421–428
Perrie, W. T. & Perry, S. V. (1970) *Biochem. J.* **119**, 31–38
Perrie, W. T., Smillie, L. B. & Perry, S. V. (1972a) *Biochem. J.* **128**, 105P–106P
Perrie, W. T., Smillie, L. B. & Perry, S. V. (1972b) *Cold Spring Harbor Symp. Quant. Biol.* **37**, 17–18
Perrie, W. T., Smillie, L. B. & Perry, S. V. (1973a) *Biochem. J.* **135**, 151–164
Perrie, W. T., Thomas, M. A. W. & Perry, S. V. (1973b) *Biochem. Soc. Trans.* **1**, 860–861
Perry, S. V. & Cole, H. A. (1973) *Biochem. J.* **131**, 425–428
Perry, S. V. & Cole, H. A. (1974) *Biochem. Soc. Trans.* **2**, 89–90
Perry, S. V. & Grey, T. C. (1956a) *Biochem. J.* **64**, 184–192
Perry, S. V. & Grey, T. C. (1956b) *Biochem. J.* **64**, 5P
Perry, S. V., Cole, H. A., Head, J. F. & Wilson, F. (1972) *Cold Spring Harbor Symp. Quant. Biol.* **37**, 251–262
Perry, S. V., Head, J. F. & Winter, M. R. C. (1973) *Abstr. Symp. Calcium-Binding Proteins, Warsaw*
Potter, J. & Gergely, J. (1972) *Fed. Proc. Fed. Amer. Soc. Exp. Biol.* **31**, Abstr. 501
Rome, E., Hirabayashi, T. & Perry, S. V. (1973) *Nature (London)* in the press
Schaub, M. C. & Perry, S. V. (1969) *Biochem. J.* **115**, 993–1004
Schaub, M. C., Hartshorne, D. J. & Perry, S. V. (1967) *Biochem. J.* **104**, 263–269
Spudich, J. A., Huxley, H. E. & Finch, J. F. (1972) *J. Mol. Biol.* **72**, 619–632
Stull, J. T., Brostrom, C. O. & Krebs, E. G. (1972) *J. Biol. Chem.* **247**, 5272–5274
Wilkinson, J. M., Perry, S. V., Cole, H. A. & Trayer, I. P. (1971) *Biochem. J.* **124**, 55P–56P
Wilkinson, J. M., Perry, S. V., Cole, H. A. & Trayer, I. P. (1972) *Biochem. J.* **127**, 215–228

Calcium Ions: Their Ligands and Their Functions

By R. J. P. WILLIAMS

Wadham College, University of Oxford, Oxford OX1 3QR, U.K.

Synopsis

Calcium cations control very different biological actions from those controlled by other cations, for example magnesium. Calcium binding to proteins is often much stronger than that of magnesium and calcium frequently acts as a cross-linking agent. This article outlines some of the properties of the calcium ion which have been found to influence the nature of its complexes with small model ligand systems and which may well give rise to the selectivity of action in biology. The importance of the structure of the co-ordination sphere and of hydration in both this first sphere and in the second sphere around the cation are stressed. As calcium is distributed in a very uneven manner in different cells and cell particles a full understanding of calcium functions also requires a knowledge of the way in which calcium concentration gradients are developed and released, for the dynamics of calcium flow are an essential part of its action.

Introduction

Calcium ions perform a number of functions in biology as has been shown by the preceding papers of this meeting. I wish to draw attention to the connexion between the properties of the calcium cation and these functions. The problem is not just one of the relationship between function and structure, for the calcium ion concentration in different parts of the cell is controlled by metabolism. Again the action of calcium can be antagonized by a variety of cations of both inorganic and organic origin.

The general situation may be inspected by an equation of the following form:

$$\text{Activity generated by calcium cations} \propto \frac{[Ca^{2+}] \cdot K_{Ca} \cdot p_{CaX}}{[M^{n+}] \cdot K_M \cdot p_{MX}}$$

Where $[Ca^{2+}]$ is the concentration of calcium ions free in a particular aqueous medium and $[M^{n+}]$ is the concentration of any antagonistic cation (organic or inorganic) in the same medium. The binding constants of the two different cations to some organic moiety, X, involved in activity are given by the different K values, these constants being measured in water, and the partitition coefficients of the complexes into a separate medium, say a membrane, are given by p. An understanding of calcium action is helped by a knowledge of the above parameters but a full appreciation can only come from the further knowledge of the structure–function relationship for the calcium complex itself. The last point also introduces the question as to why a second cation which binds X antagonizes rather than mimics the function of calcium. The equation as it stands applies to equilibrium conditions and I shall examine the terms in the above equation in the first instance under resting stationary-state conditions of a cell, which can be thought of as in local equilibrium, turning later to the dynamic situation in which the flow of calcium ions from one cell compartment to another generates activity.

Calcium Concentration

Calcium concentration varies from $0.1\,\mu M$, in some fresh waters to $1\,mM$ in others. It is quite high in the sea. Inside the cytoplasm of cells calcium concentration is usually $0.1\,\mu M$ or lower whereas in the bloodstream it is greater than $1\,mM$. There are also special cell structures that accumulate and buffer calcium concentrations. For example, the sarcoplasmic reticulum must have a concentration of calcium above $0.1\,mM$ and it also has a great deal of calcium sequestering protein which can bind as many as 30 calcium atoms/mol. It is obviously essential to state where a given biological molecule is in relation to the calcium concentration before discussing the role it plays in calcium binding. Equally important is the concentration, again under metabolic control, of the ions M^{n+}.

Binding Strength

The main binding groups for calcium are oxyanions. The binding strength of a cation such as calcium is known to vary over at least ten orders of magnitude in model complexes containing from one to four such residues, e.g. carboxylates. In biological systems the binding constant to some proteins in cells such as the carp albumin is at least 10^7, four carboxylate residues are involved, whereas binding to proteins such as concanavalin (two carboxylates) is of smaller stability (Table 1). Binding to sugar carboxylates, phosphates and sulphonates will also be closely related to the number of anionic groups available. Of course, the binding constants to molecules inside cells must be higher than $1 \times 10^{-6}\,M^{-1}\cdot l$ for the $[Ca^{2+}]$ is less than $1\,\mu M$. In the bloodstream of animals this binding strength is not required, $[Ca^{2+}] = 1\,mM$.

The disposition of organic-molecule-binding sites, both outside and inside biological cells, is such that the calcium ion concentration helps to control the steady-state structures of cell membranes and walls, cell–cell conglomerates and structural units inside the cell. Now it seems that the magnesium cation par-

Table 1. *Binding sites of proteins for calcium*

Protein	Site	Binding character
Carp albumin	Four carboxylate residues	$K > 10^6$
Thermolysin	Three carboxylate residues	$K > 10^4$
Bacterial nuclease	Three carboxylate residues	$K > 10^4$
Concanavalin	Two carboxylate residues	$K \geqslant 10^2$
Ionophore* X-537A	One carboxylate residue	K is probably very low but the ionophore is membrane soluble
Lysozyme	Two carboxylate residues	$K > 10^1$

* Note that the present paper has not described binding selectivity based on the ring (hole) size of a ligand as this aspect is described elsewhere (see Williams, 1970, 1972).

Table 2. *Some typical calcium salt structures*

Note that when calcium is six-co-ordinate the structures may be more regular.

Salt	Ca(II) co-ordination no.	Ca–O distances (Min.)	(Max.)	Reference
$CaHPO_4,2H_2O$	8	2.44	2.82	Beevers (1958)
$Ca(H_2PO_4)_2,H_2O$	8	2.30	2.74	MacLennan & Beevers (1956)
Calcium 1,3-diphosphoryl-	6	2.26	2.36	Beard & Lenhert (1968)
imidazole	7	2.27	2.78	Beard & Lenhert (1968)
Calcium dipicolinate, $3H_2O$	8	2.36	2.57	Strahs & Dickerson (1968)
$CaNa(H_2PO_2)_3$	6	2.31	2.33	Matsuzaki & Iitaka (1969)
Calcium tartrate, $4H_2O$	8	2.39	2.54	Ambady (1968)
$Ca(C_6H_9O_7)_2,2H_2O$	8	2.39	2.47	Belchin & Carlisle (1965)

ticularly antagonizes these functions. We must look at differences in the concentration and in the binding strengths of the two cations.

The magnesium ion has a free ion concentration of about 1 mM both outside and inside the cell. However, it binds less well than calcium to multi-dentate ligands of carboxylate donors such as EDTA and EGTA [ethanedioxybis(ethylamine)tetra-acetic acid] (Williams, 1970). Thus some of the proteins of Table 1 are designed as calcium-binding proteins. If a protein inside a cell is to bind calcium the discrimination must be much larger between the magnesium- and calcium-binding strengths, than outside the cell. Clearly, the selective binding of magnesium or calcium can be very carefully controlled through a combination of concentration and binding-strength differences.

Binding strength is not just a matter of the numbers of charged groups, however, for if it were so then the problem of the competition between different cations would be easily described. Cations such as sodium would bind less well than calcium and calcium would bind less well than lanthanum. However, it is known that this is not universally the case. Notice that all these cations have exactly the same size, approximately 0.1 nm (1.0 Å) radius, so that it is not charge/size ratio that controls binding strengths either. Examination of crystal structures of complicated hydrated salts (Table 2) shows that additional factors arise from second sphere, not co-ordination sphere, interactions. Amongst these are the dielectric of the medium, the structure and strength of the hydrogen bonding to the water molecules of hydration and the other ligands. Data on partition and on the solubility of salts indicate that such factors can lead to quite unexpected selectivity for one cation rather than another. Note that $CaSO_4$ is more insoluble than either Na_2SO_4 or $La_2(SO_4)_3$. It may well be that we have concentrated too much attention upon the immediate environment of the cations. In a protein this could be particularly hazardous as the protein may well adjust its structure to compensate for excess of charge at a given site. Clearly four carboxylate groups repel one another. The degree to which sodium, calcium, or lanthanum decrease this repulsion will govern the conformation of the protein. There is no detailed study of how cation-binding changes conformation of a protein and hence alters energetic and functional properties of the cation binding.

So far we have considered only the binding of a single ligand to a single metal cation. It is obviously necessary to extend the discussion to more complicated situations where there is more than one binding group and/or metal cation. Such a situation occurs in the formation of bone, shell and similar organized biological structures and to understand these we must look at the nature of co-operative interactions and cross-linkings.

Co-operative Interactions

In a solid such as calcium oxide, which has the NaCl structure, the lattice energy is nearly twice as big as the energy of the gaseous diatomic dipole $Ca^{2+}O^{2-}$ owing to the co-operative interaction of the dipoles in the lattice. Such co-operative energies arise to a lesser degree in all highly dipolar assemblies, e.g. water. Obviously they are also important in controlling the structures and functions of more irregular phases such as bones, shells, soils etc. It is a feature of the latter phases (they are not simple compounds) that they are composed of two or three major components and several minor components. The major components usually include the calcium ion and some anion of either different size or different charge type from calcium, e.g. phosphate. Even with a pair of such components it is not necessarily true that the most stable structure that arises will be regular. Again there are often several crystal forms of comparable stability, e.g. calcite and aragonite. The addition of the minor components then permits a variety of structures to develop with a variety of compositions. In biology one of these minor components is protein and it could be that the structures of many biological phases such as bone and shell are controlled by the precise protein in them. Initially we must look at the structures of individual cation compounds, though ultimately we wish to find out how biology turns the co-operativity to use much as we begin to see how it utilizes the binding energy of single molecule sites for a single cation.

Structure

There are surprisingly few constraints upon the structures of solid calcium salts (Table 2). Bond distances are very flexible, bond angles can be equally adjustable, co-ordination number can vary from six to ten. Magnesium has a much more restricted range of structures. Calcium is therefore a cation that will almost invariably displace magnesium from a binding site composed of a complex set of multi-dentate and flexible ligands. (The only general way to protect magnesium sites of this kind from calcium is to pump calcium out!) It is often the case that these complexes will retain residual hydration of the calcium ion. Now this flexibility in the calcium ion co-ordination sphere will permit a great variety of co-operative packings. For example, cross-linking between organic groups may be achieved with the calcium ion whereas no such cross-linking may be found with magnesium. We may ask, for example, is this why calcium is so important in exocytosis? Action here is not just a matter of charge neutralization ($Mg^{2+} = Ca^{2+}$) but a matter of binding together two membrane surfaces. Magnesium could antagonize this action of calcium for it will not cross-link although it may bind equally well to any single group or surface. For example, magnesium ions bind as well as calcium to single phosphate, oxalate, carbonate and sulphate anions but co-operative energies are much greater in the calcium salts and therefore the calcium salts are much more insoluble than the corresponding magnesium compounds. This could be the basis of many biological phenomena and could also be the reason that man uses calcium salts in building materials such as plasters and concretes. Again in so far as calcium cross-links phospholipids in the surface of membranes it creates asymmetry in them and therefore energizes the membranes through the differential calcium concentration across the membrane (Williams, 1972). Magnesium ions modulate these effects.

Dynamic Aspects

There are two ways in which the steady state can be adjusted; (1) organic matter can flow to calcium, as in digestion, (2) calcium ions can flow to the organic centres, as in muscle action, down respective gradients. We need to ask, how are such gradients maintained for they require a constant energy input?

The ability of ATP to phosphorylate proteins introduces one direct way of connecting metabolism to calcium pumping. Thus, if we suppose that a given protein has a binding constant of 10^5 owing to three carboxylate groups and that the binding site has a neighbouring serine, it is easily seen that the phosphorylation of the serine could lead to a binding constant of 10^7. Calcium transport could be readily coupled to phosphate transport, if the protein were to be phosphorylated at one place and dephosphorylated at another. On the other hand, there are direct ways of transferring calcium unrelated to phosphorylation of proteins.

It is a well recognized feature of calcium compounds that they can form different hydrates with rather closely similar stabilities. This arises from the ease with which calcium alters its co-ordination number and bond distances (see above). The variable hydration could easily arise in calcium–protein complexes so that there could be two hydrates of closely related energy. It is already generally realized that the formation of anhydrides between phosphate groups is a major store of energy in ATP and that dehydration is the main method of making unstable biological polymers. The different hydration states of cations also permit the dehydration reaction to be used to energize the complexes of inorganic cations such as calcium. The dehydration could be coupled to a change of the protein structure to which the calcium was bound. Normally the rapid exchange of water from around a calcium ion would not make this energy store very much more than a transient phenomenon. In the event that calcium was trapped in a hydrophobic region of a protein where access of water was less facile, then the dehydration and consequent change of protein geometry could generate a calcium protein in a high-energy form of some persistence. Similar considerations apply to other cations and it is a simple matter to go forward to discuss the transport of cations across membranes relying upon these dehydration/hydration equilibria to control energized transport.

Given that energy can be put into a system so as to generate calcium gradients across membranes, and that differential interaction of calcium with the two sides of the membranes and with cell constituents will therefore energize the whole of the cell's structures, the possibility arises of evolving control mechanisms. Action can arise from a direct release of control over the cation steady state when the ions diffuse down their concentration gradients. The return to the initial resting state, which is often called relaxation but is really re-energization, is the regeneration of the steady state of high energy. However, because of the presence of the membranes there are other effects for which we must watch. Binding of groups (hormones or drugs) that remove the membrane-bound calcium could allow the membrane to change to a new structure. This change of structure could then be a trigger of further changes. Return to the steady state requires the removal of the drug by metabolism. Clearly a very subtle relationship exists between calcium ion concentration and various energized structures in cells and there are several ways in which calcium ion concentrations could be adjusted (Lehninger, 1970).

An Overall View

The resting state of a biological system is, as far as calcium ion concentration is concerned, very far from the equilibrium situation. This is also true to a lesser degree for many other cations. In this resting state the removal of calcium from within the cell allows the stabilization on a temporary basis of many thermodynamically unstable structures both of proteins and vesicular membranes. Outside the cell where the calcium ion concentration is higher the calcium maintains major structural units on a much longer time-scale. Perturbations of the system will not affect the external regions grossly as the calcium concentration is high enough to buffer this medium fairly effectively. Inside the cell, however, a switch in concentration permits calcium to act almost like a hormone switching cell structures. Observed 'effects' as calcium enters a cell have been analysed in many of the papers at this meeting. In the present paper I have tried to show why it is calcium and no other cation that performs this role. By way of distinction the role of magnesium is to maintain internal cell arrangements on a longer time-scale. The magnesium concentration gradients are buffered by energy utilization in a totally different concentration range. The niceties of the control of these two cations are just one more example of the exploitation of chemical differences in evolution.

References

Ambady, G. K. (1968) *Acta Crystallogr.* **24B**, 1548–1557
Beard, L. N. & Lenhert, P. G. (1968) *Acta Crystallogr.* **24B**, 1529–1539
Beevers, C. A. (1958) *Acta Crystallogr.* **11**, 273–277
Belchin, A. A. & Carlisle, C. H. (1965) *Acta Crystallogr.* **19**, 103–111
Lehninger, A. L. (1970) *Biochem. J.* **119**, 129–138
MacLennan, G. & Beevers, C. A. (1956) *Acta Crystallogr.* **9**, 187–190
Matsuzaki, T. & Iitaka, Y. (1969) *Acta Crystallogr.* **25B**, 1932–1938
Strahs, G. & Dickerson, R. E. (1968) *Acta Crystallogr.* **24B**, 571–578
Williams, R. J. P. (1970) *Quart. Rev. Chem. Soc.* **24**, 331–365
Williams, R. J. P. (1972) *Physiol. Chem. Phys.* **4**, 427–435

Author Index

Numbers in italics are pages where references are listed at the end of each article.

A

Abbott, B. C., 45, *49*
Adelstein, R. S., 130, *131*
Ahmed, K., 106, *108*
Ambady, G. K., 135, *138*
Amsterdam, A., 20, *26*
Anderson, P., 16, 17, 19, *28*
Antoniw, J. F., 61, 62, 63, *72*
Argent, B. E., 17, *26*
Arienti, G., 103, *109*
Ash, J. F., 24, *28*
Ashby, C. D., 56, *72, 73*
Ashley, C. C., 31, 32, 33, 34, 35, 37, 38, 39, 40, 41, 42, 43, 44, 45, 46, 47, 48, *49*
Azzi, A., 90, *107, 109*
Azzone, G. F., 90, 91, *109*

B

Bailey, C., 66, *72*
Baker, P., 5, 6, 14, 15, 23, *26*
Baker, P. F., 32, 36, 40, 41, *49*
Balcavage, W. X., 93, *107*
Banks, P., 3, 5, 20, *26*
Bargmann, W., 9, 11, *26, 27*
Barker, M. D., 93, *108*
Barrera, C. R., 76, *88*
Barrett, A. J., 55, *72*
Barylko, B., 30, *49*, 118, *131*
Baskin, R. J., 33, *49*
Batra, A., 102, *107*
Baum, H., 104, *108*
Beard, L. N., 135, *138*
Becker, G., 94, 95, *108*
Becker, K., 9, *27*
Beckman, E., 71, *72*
Beechey, R. B., 93, *108*
Beevers, C. A., 135, *138*
Belchin, A. A., 135, *138*
Benedeczky, I., 6, 19, *26*
Bennett, H. S., 1, 22, *26*
Berl, S., 24, 25, *26, 27*
Bernstein, J., 24, *28*
Berther, J., 56, *72*
Biason, M. G., 103, *109*
Biggins, R., 5, *26*
Birks, R. I., 13, *26*
Bishop, J. S., 69, 70, *72*
Bishop, R., 5, *26*

Blaustein, M. P., 14, *26*, 36, *49*
Blinks, J. R., 41, 42, *49, 50*
Bloom, G. D., 16, *26*
Bloom, S., 106, 107, *107*
Borle, A. B., 96, *107*
Bradley, M. P., 24, *28*
Bratvold, G. E., 61, 65, *72*
Brekke, C., 30, *49*, 127, *132*
Bremel, R. D., 46, *50*, 126, *131*
Bridges, T. E., 9, *28*
Briggs, F. N., 101, *108*
Briskey, E. J., 112, *113*
Brostrom, C. O., 53, 56, 57, 61, 63, 66, 70, *72, 73*, 127, *132*
Brown, D. H., 51, *72*
Brown, N. B., 69, 70, *72*
Buchholz, M., 93, *108*
Buchtal, F., 124, *131*
Bunt, A. H., 11, *26*
Burgett, M. W., 76, 77, 83, *88*
Burston, J., 106, *109*
Burton, A. M., 9, *26*
Bygrave, F. L., 103, *107, 108*

C

Caldwell, P. C., 30, 31, 32, 33, 34, 35, *49*
Calkins, D., 56, *73*
Cameron, D. A., 104, *107*
Cancilla, P. A., 106, 107, *107*
Capony, J. P., 116, *132*
Carafoli, E., 85, *88*, 89, 90, 91, 92, 93, 94, 95, 96, 98, 102, 103, 104, 106, 107, *107, 108, 109*
Carlisle, C. H., 135, *138*
Carlsen, F., 124, *131*
Case, R. M., 17, 18, *26*
Cassens, R. G., 112, *113*
Caulfield, J. B., 106, 107, *108*
Ceccarelli, B., 12, *26*
Ceccarelli, D., 104, *108*
Chad, I., 20, *26*
Chakravarty, N., 16, *26*
Chance, B., 42, *50*, 102, *108*
Chappell, J. B., 93, 103, *108*
Chen, A. C. N., 37, *49*
Cheng, K. W., 9, *26*
Chiga, M., 106, *108*
Christian, B., 5, *26*

Chubb, I. W., 12, 13, *26*
Clausen, T., 18, *26*
Clementi, F., 20, *26*
Coceani, F., 104, *108*
Cochrane, D. E., 17, 25, *26*, *27*
Cohen, P., 52, 53, 54, 57, 58, 59, 61, 62, 63, 64, 65, 67, 68, 70, *72*, 116, *131*
Cohen, P. T. W., 70, *72*
Cole, H. A., 30, *50*, 66, *72*, 118, 123, 124, 126, 127, *132*
Collier, B., 25, *28*
Conti, M. A., 130, *131*
Cooper, R. A., 53, 56, 67, *72*
Coore, H. G., 77, 86, *88*
Corbin, J. D., 56, 70, *72*
Cori, C. F., 51, 52, 67, *72*
Cori, G. T., 52, *72*
Crang, R. E., 105, *108*
Crovetti, F., 104, *108*
Currie, N., 5, *26*

D

Dabrowska, R., 118, 127, *131*
D'Agostino, A. N., 106, 107, *108*
Danforth, W. H., 52, 70, 67, *72*
Dawson, A. P., 94, *109*
Dean, P. M., 17, *26*
de Bernard, B., 94, 95, *108*, *109*
de Duve, C., 2, *26*
Delange, R. J., 53, 56, 61, 67, *72*, 129, *132*
Del Castillo, J., 13, *26*
Denton, R. M., 75, 77, 78, 79, 84, 85, 86, *88*, 103, *108*
De Potter, W. P., 12, 13, *26*
De Robertis, E. D. P., 1, 2, *26*
De Schaepdryver, A., 12, 13, *26*
De Virgilis, G., 20, *26*
Diament, B., 17, *26*
Dickerson, R. E., 135, *138*
Diner, O., 3, 6, 19, *26*
Dingle, J. T., 55, *72*
Donnellan, J. F., 93, *108*
Douglas, W. W., 2, 3, 4, 5, 6, 7, 8, 9, 10, 11, 12, 13, 14, 15, 16, 17, 20, 21, 22, 23, 24, 25, *26*, *27*
Drabikowski, W., 30, *49*, 118, 127, *131*
Drabowska, R., 30, *49*
Drahota, Z., 98, *108*
Draper, M. H., 124, *131*
Dreifuss, J. J., 9, *27*
Drummond, G. I., 60, *72*
Duewer, T., 52, 70, *72*
Dunnett, S. J., 94, *109*
Dydynska, M., 127, *131*

E

Ebashi, F., 101, *108*, 118, *131*
Ebashi, S., 30, 36, *49*, *50*, 53, 66, *72*, 101, *108*, 118, 124, *131*
Eley, M. H., 76, *88*
Ellory, J. C., 31, 32, 37, 40, *49*
Endo, M., 30, 48, *49*, 53, 66, *72*, 101, *108*
England, P. J., 66, 70, *72*, 127, *132*
Erlichman, J., 56, *73*

F

Farquhar, M. G., 2, 20, *27*
Fatt, P., 31, 33, *49*
Fawcett, C. P., 8, 9, *27*
Feinstein, M. B., 3, 5, *28*
Felton, S. P., 37, *49*
Finch, J. F., 124, *132*
Fink, C. J., 24, *27*
Fischer, E. H., 52, 53, 56, 60, 64, 65, 69, 70, *72*, *73*
Ford, L. E., 48, *49*
Forsling, M., 9, *26*
Fosset, M., 65, *72*
Franchimont, P., 9, *27*
Fried, M., 65, *72*
Friesen, H. G., 9, *26*
Fuchs, F., 101, *108*
Funcke, H. V., 77, *88*

G

Gallagher, C. H., 106, *108*
Gamble, R. L., 94, 98, *107*, *108*
Gammeltoft, S., 86, *88*
Garland, P. B., 75, *88*
Gazzotti, P., 85, *88*, 90, 91, 93, 94, 95, 98, *107*, *108*, *109*
Gergely, J., 30, *49*, 118, 120, 127, *132*
Geschwind, I. I., 8, 17, *27*
Gliemann, J., 86, *88*
Glinsmann, W. H., 63, *72*
Goldberg, N. D., 69, 70, *72*
Gomez-Puyou, A., 94, 95, *108*
Gonzales, F., 105, *108*
Gonzalez, C., 56, *73*
Graves, D. J., 65, 71, *72*, *73*
Grau, J. D., 9, *27*
Greaser, M. L., 30, *49*, 112, *113*, 118, 127, *132*
Green, N. M., 113, *113*
Greenawalt, J. W., 104, *108*
Grey, T. C., 117, *132*
Grisham, J. W., 2, *28*
Gupta, D. N., 106, *108*

AUTHOR INDEX

H

Hackney, J. H., 91, *109*
Hainaut, K., 32, *49*
Hall, B. V., 106, *109*
Hall, C. E., 124, *132*
Hamilton, L., 76, *88*
Hamoir, G., 116, *132*
Hansford, R. G., 67, 72, 93, 103, *107*, *108*
Hanson, J., 124, *132*
Hardwicke, P. M. D., 113, *113*
Harigaya, S., 36, *49*
Hartshorne, D. J., 117, 118, 120, *132*
Harvey, A. M., 13, *27*
Harwood, J. P., 60, *72*
Haschke, R. H., 53, 60, 69, 70, *72*
Hastings, J. W., 41, 42, *49*
Hava, M., 6, *28*
Hayakawa, T., 53, 55, 71, *72*, 116, *132*
Head, J. F., 30, *50*, 120, 121, 122, 123, 124, 125, 126, 127, *132*
Heggtveit, A. H., 106, 107, *108*
Heilbrunn, L. V., 29, *49*
Heilmeyer, L. M. G., 53, 60, 66, 69, 70, *72*
Hellam, D. C., 45, 48, *49*
Helle, K., 3, *26*
Helmreich, E., 52, 67, *72*
Hendrick, J. C., 9, *27*
Herdson, P. B., 106, *108*
Herman, L., 106, 107, *108*
Herz, R., 45, 46, *50*
Heuser, J. E., 11, 12, *27*
Hickinbottom, J. P., 59, 60, *73*
Hill, A. V., 30, *49*
Hirabayashi, T., 124, 127, *132*
Hirsch, J. G., 25, *28*
Hitt, J. B., 105, *108*
Ho, E. S., 56, 57, 63, *73*
Hodge, A. J., 124, *131*
Hodgkin, A. L., 13, *27*, 36, 40, 41, *49*
Hoeckstra, W. G., 112, *113*
Hokin, L. E., 17, *27*
Holsen, R. C., 105, *108*
Homan, W., 105, *108*
Hope, D. B., 9, *28*
Horsfield, G., 17, 19, 20, *27*
Horton, E. W., 104, *108*
Horwitz, R. A., 71, *72*
Hosoi, K., 53, *72*
Howell, S. L., 24, *27*
Hoyle, G., 30, 33, 45, *49*
Hubbard, J. I., 8, 12, 13, 14, 15, *27*
Hucho, F., 76, 77, 78, 83, *88*
Huijing, F., 69, 70, *72*
Hunkeler, F. L., 53, 56, 59, 60, 61, *72*, *73*
Hurd, S. S., 70, *72*
Hurlbut, W. P., 12, *26*
Huston, R. B., 52, 64, *72*
Huxley, H. E., 124, *132*

I

Iitaka, Y., 135, *138*
Iles, G. H., 111, 112, *113*, 116, *132*
Illingworth, B., 51, *72*
Inesi, G., 36, *49*
Izumi, F., 24, *27*
Izutsu, K. T., 37, *49*

J

Jaanus, S. D., 3, 5, *28*
Jakus, M. A., 124, *132*
Jennings, R. B., 106, *108*, *109*
Jöbsis, F. F., 36, 37, *49*
Johnson, F. H., 32, 37, 38, *50*
Judah, J. D., 106, *108*
Julian, F. J., 45, 48, *50*
Jungas, R. L., 77, *88*

K

Kadota, K., 22, *27*
Kanaseki, T., 22, *27*
Kanno, T., 3, 5, 10, 13, 16, 17, 21, 25, *26*, *27*
Karnovsky, M. J., 105, *108*
Kashimoto, T., 24, *27*
Katz, B., 13, 14, 20, *26*, *27*, 31, 33, *49*
Kay, C. M., 118, 120, *132*
Keller, P. J., 65, *72*
Kemp, R. G., 53, 56, 67, *72*, 129, *132*
Kendrick-Jones, J., 30, 45, 46, *50*, 127, 128, 129, *132*
Keynes, R. D., 13, *27*
Kimura, S., 103, *108*
King, C. A., 56, 70, *72*
Kirshner, G., 6, *27*
Kirshner, N., 6, *27*
Kirtland, S. J., 104, *108*
Kletzien, R. F., 25, *27*
Klingenberg, M., 93, *108*
Knappeis, C. G., 124, *131*
Kodama, A., 101, *108*, 118, *131*
Kondo, K., 103, *108*
Konosu, S., 116, *132*
Krebs, E. G., 52, 53, 55, 56, 57, 59, 60, 61, 63, 64, 65, 66, 67, 70, 71, *72*, *73*, 116, 127, 129, *132*
Krisch, B., 9, *27*
Krüger, P. G., 17, *26*

L

Lacy, P. E., 2, 24, *27, 28*
Lafferty, F. W., 106, *108*
Laguens, R., 102, *108*
Lagunoff, D., 16, *27*
Langer, G. A., 104, 105, *108*
Larner, J., 51, 69, 70, *72*
Laschi, R., 106, 107, *108*
Lastowecka, A., 25, *28*
Lederis, K., 9, *28*
Le Douarin, N., 20, *27*
Legato, M. J., 104, 105, *108*
Legros, J. J., 9, *27*
Lehman, W., 30, 45, 46, *50*, 127, 128, 129, *132*
Lehninger, A. L., 89, 90, 91, 92, 93, 94, 95, 96, 98, 102, 103, 104, 105, *107*, *108*, *109*, 138, *138*
Le Lievre, C., 20, *27*
Lenhert, P. G., 135, *138*
Linn, T. C., 75, 76, 77, 78, *88*, 129, *132*
Livett, B. G., 9, *28*
Löffler, G., 77, *88*
Loschen, G., 42, *50*
Love, D. S., 65, *72*
Lowe, A. G., 31, 32, 33, 34, 35, *49*
Lowenstein, W., 6, *27*
Luduena, M. A., 24, *28*
Lugli, G., 103, *108*
Lyon, J. B., 70, *72*
Lywood, D. W., 13, *27*

M

MacIntosh, F. C., 13, *26, 27*
MacLennan, D. H., 111, 112, *113*, 116, *132*
MacLennan, G., 135, *138*
Marchalonis, J. J., 112, *113*
Martin, B. R., 75, 77, 78, 79, 84, 85, 86, *88*, 103, *108*
Martin, J. H., 105, *108*
Martin, M. J., 9, *26*
Martonosi, A., 112, *113*
Masaki, T., 124, *132*
Matsuzaki, T., 135, *138*
Matthews, E. K., 9, 17, 20, *26, 27*
Matthews, J. L., 105, *108*
Mattingly, P. H., 41, 42, *49*
Mattoon, J. R., 93, *107*
Mauro, A., 12, *26*
Mayer, W. L., 65, *72*
McLean, A. E. M., 106, *108*
McNeil, P. A., 33, *49*
McPherson, M. A., 25, *27*
Mela, L., 90, 93, *108*
Meldolesi, J., 20, *26*
Meli, J., 103, *108*
Meyer, F., 53, 60, 69, *72*
Meyer, S. E., 60, *72*
Meyer, W. L., 52, 64, *72*
Mihashi, K., 124, *131*
Mikiten, T. M., 14, *27*
Miledi, R., 14, *27*
Mishra, R. R., 106, 107, *108*
Mitchell, G., 41, 42, *49*
Mitchell, P., 94, *108*
Mizel, S. B., 25, *27*
Moisescu, D. G., 39, 41, 42, 43, 44, 45, 46, 47, 48, *49*
Mongar, J. L., 16, *27*
Moore, C., 90, *108*
Mueller, H., 118, *132*
Muir, L. W., 65, *72*
Munk, P., 76, *88*
Murphy, J. V., 89, *109*
Murray, A. C., 118, 120, *132*

N

Nagasawa, J., 6, 7, 9, 10, 11, 12, 20, 21, 22, 23, *26, 27*
Nakazato, Y., 25, *27*
Namihara, G., 76, *88*
Nicklas, W. J., 24, 25, *26, 27*
Nonomura, Y., 124, *132*
Nordman, J. J., 9, *27*
Normann, T. C., 9, 11, *27*
Nowak, E., 30, *49*
Nowinski, W. W., 1, *26*

O

O'Connor, M. J., 36, 37, *49*
Ogata, E., 103, *108*
Ogawa, Y., 36, 48, *49*
Ohnishi, T., 36, *50*
Ohtsuki, I., 30, *49*, 53, 66, *72*, 124, *132*
Oka, M., 24, *27*
Østerlind, K., 86, *88*
Ozawa, E., 53, *72*

P

Pace-Asciak, C., 104, *108*
Page, S. G., 124, *132*
Paimre, M., 3, 5, *28*
Palade, G., 2, *27*
Palay, S. L., 10, *27*
Panfili, E., 94, 95, *108, 109*
Paschall, H. A., 104, *107*
Pask, H. T., 77, 78, 84, 85, *88*
Patriarca, P., 96, 98, *108*

Peachey, L. D., 104, 105, *108*
Pearse, A. G. E., 20, *28*
Pearson, E. S., 106, *108*
Pechere, J. F., 116, *132*
Pelley, J. W., 76, 77, 78, 83, *88*
Perdue, J. F., 25, *27*
Perkins, J. P., 53, 55, 56, 57, 59, 63, *72, 73*, 116, *132*
Perrie, W. T., 66, *72*, 129, 130, *132*
Perry, S. V., 30, *50*, 66, *72*, 117, 118, 120, 121, 122, 123, 124, 125, 126, 127, 129, 130, *132*
Pettit, F. H., 75, 76, 77, 78, 81, *88*, 129, *132*
Pirotta, M., 103, *109*
Podolsky, R. J., 45, 48, *49*
Poisner, A. M., 3, 5, 6, 13, 14, 17, 18, 24, *26, 27, 28*
Polak, J. M., 20, *28*
Porcellati, G., 103, *109*
Portenhauser, R., 77, *88*
Porter, J., 70, *72*
Porter, K. R., 22, *28*
Portzehl, H., 32, *50*
Posner, J. B., 60, *72*
Potter, J., 30, *49*, 120, 126, *132*
Powell, A. E., 8, 9, *27*
Powell, C. A., 69, *72*
Pratje, E., 66, *72*
Puszkin, S., 24, 25, *26, 27*
Pysh, J. J., 12, *28*
Pyun, H. Y., 120, *132*

R

Racker, E., 111, 112, *113*
Rall, T. W., 56, *72*
Randall, D. D., 76, 77, 78, 83, *88*
Randle, P. J., 75, 77, 78, 79, 84, 85, 86, *88*, 103, *108*
Reed, L. J., 75, 76, 77, 78, 81, 83, *88*
Reed, P. W., 98, *109*
Rees, K. R., 106, *108*
Reese, T. S., 11, 12, *27*
Reid, L. J., 129, *132*
Reimann, E. M., 56, 59, 60, *72, 73*
Reiss, I., 45, 46, *50*
Reynafarje, B., 91, 93, *109*
Reynolds, E. S., 106, 107, *108, 109*
Ridgway, A. L., 36, 40, 41, *49*
Ridgway, E. B., 38, 39, 43, *49, 50*
Riley, W. D., 53, 56, 61, 67, *72*, 129, *132*
Ringer, S., 29, *50*
Robinson, R. A., 104, *107*
Roche, T. E., 75, 76, 77, 81, 83, *88*
Röhlich, P., 16, 17, 19, *28*

Rome, E., 124, *132*
Ronchetti, I., 106, 107, *108*
Rose, R. M., 45, 48, *49*
Rosen, O. M., 56, *73*
Rossi, C. S., 89, 90, 94, 95, 96, 98, 102, 103, 104, *107, 108, 109*
Roth, T. F., 22, *28*
Rubin, C. S., 56, *73*
Rubin, R. P., 2, 3, 4, 5, 6, 8, 13, 14, 15, 16, 23, *27, 28*
Rüdel, R., 46, *50*
Rüegg, J. C., 32, *50*
Ruffolo, P. R., 106, 107, *109*
Ryan, J. W., 6, *28*
Ryden, L., 116, *132*

S

Saari, J. C., 65, *72*
Sabatini, D. D., 2, *26*
Sachs, H., 8, 9, 11, *27*
Sacktor, B., 67, *72*, 93, 94, *107*
Saez, F. A., 1, *26*
Saiga, Y., 32, 37, *50*
Saltini, C., 94, 95, *108*
Sampson, S. R., 5, 10, 13, 21, *27*
Sandow, A., 30, *50*
Sandri, G., 94, 95, *108, 109*
Santolaya, R. C., 9, *28*
Sasko, H., 69, 70, *72*
Scarpa, A., 36, *49*, 91, *109*
Scarpelli, D. G., 106, 107, *109*
Scharrer, B., 9, *28*
Schaub, M. C., 117, 118, *132*
Schild, H., 17, *28*
Schild, H. O., 16, *27*
Schmitt, F. O., 124, *132*
Schofield, J. G., 25, *27*
Schraer, H., 105, *108*
Schrag, B. A., 106, 107, *108*
Schramm, M., 17, 20, *28*
Schulz, R. A., 9, 10, 11, 12, *27*
Schwartz, A., 36, *49*
Scratcherd, T., 17, *26*
Seeman, P., 111, 112, *113*, 116, *132*
Sell, D. E., 106, 107, *109*
Selverston, A. I., 33, *49, 50*
Selwyn, M. J., 94, *109*
Sheff, M. F., 106, *109*
Shen, A. C., 106, *109*
Shimomura, O., 32, 37, 38, *50*
Siegel, I. A., 37, *49*
Siess, E. A., 75, 78, *88*
Simon, J., 106, *109*
Smillie, L. B., 66, *72*, 129, *132*

Smith, A. D., 6, 8, 11, 12, 19, *26*, *28*
Smith, D. S., 6, 9, *28*
Smith, R. K., 18, *26*
Smith, U., 6, 9, *28*
Smyth, T., 30, *49*
Soderling, T. R., 59, 60, *73*
Sommers, H. M., 106, *108*
Sorimachi, M., 15, 25, *27*
Sottocasa, G. L., 94, 95, *108*, *109*
Spiro, D., 104, 105, *108*
Spooner, B. S., 24, *28*
Springer, A., 25, *27*
Spudich, J. A., 124, *132*
Stern, D., 25, *28*
Stern, R., 60, *72*
Strahs, G., 135, *138*
Straub, R. W., 13, *27*
Stull, J. T., 66, 70, *72*, *73*, 127, *132*
Sugita, H., 36, *49*
Sutherland, E. W., 56, *72*
Szent-Gyorgyi, A., 127, 128, 129, *132*
Szent-Györgyi, A. G., 30, 45, 46, *50*
Szpacenko, A., 127, *131*

T

Tanaka, M., 48, *49*
Taylor, E. L., 24, *28*
Taylor, S. I., 77, *88*
Taylor, S. R., 46, *50*
Teller, D. C., 70, *72*
Tessmer, G., 65, *73*
Thiers, R. E., 106, 107, *109*
Thomas, M. A. W., 130, *132*
Thon, I., 17, *28*
Thorley-Lawson, D. A., 113, *113*
Tiozzo, R., 85, *88*, 90, 91, 96, 103, 106, 107, *107*, *108*, *109*
Trayer, I. P., 118, *132*
Trayser, K. A., 65, *72*
Trifaró, J. M., 24, 25, *28*
Tuena, M., 94, 95, *108*

U

Ueda, Y., 17, 18, 25, *27*
Uttenthal, L. O., 9, *28*
Uvnäs, B., 16, 17, 19, *28*

V

Vallee, B. L., 106, 107, *109*
van Leeuwen, M., 41, 42, *50*
van Leeuwen, N., 41, 42, *49*

Van Vleet, J. F., 106, *109*
Vasington, F. D., 85, *88*, 89, 90, 91, 94, 95, 98, *108*, *109*
Vaz Ferreira, A., 2, *26*
Villar-Palasi, C., 66, 69, 70, *72*
Vinten, J., 86, *88*
Viveros, O. H., 6, *27*

W

Wakabayashi, T., 118, *131*
Walsh, D. A., 53, 55, 56, 57, 59, 60, 63, *72*, *73*, 116, *132*
Walster, G. E., 30, *49*
Weber, A., 45, 46, *50*, 102, *109*, 16, *131*
Weitzman, M., 9, *28*
Wenger, J. I., 69, 70, *72*
Wenner, C. E., 91, *109*
Wessells, N. K., 24, *28*
Whitehouse, S., 77, *88*
Wieland, O., 75, 77, 78, *88*
Wiercinski, F. J., 29, *49*
Wiley, R. G., 12, *28*
Wilkinson, J. M., 118, *132*
Williams, R. J. P., 134, 135, 137, *138*
Williamson, J. R., 2, *28*
Wilson, F., 123, 124, 126, *132*
Wilson, F. J., 30, *50*
Wilson, L., 25, *27*
Winkler, H., 6, *28*
Winter, M. R. C., 120, 125, *132*
Wolfe, L. S., 104, *108*
Wong, F. T. S., 116, *132*
Wood, J., 93, *108*
Woodhouse, M. A., 106, *109*
Wosilait, W. D., 56, *72*
Wotter, J., 127, *132*
Wren, J. R., 24, *28*

Y

Yamada, K. M., 24, *28*
Yamaguchi, M., 30, *49* 127, *132*
Yip, C. C., 111, 112, *113*, 116, *132*
Yoda, W. T., 37, *49*
Yoshitoshy, Y., 103, *108*
Young, D. A., 24, *27*

Z

Zacks, S. I., 106, *109*
Zieve, F. J., 63, *73*
Zigmond, S. H., 25, *28*

Subject Index

A

Acetylcholine,
  calcium and effect of, 2, 3, 5, 6, 18
  calcium-dependent release of, 13, 14
Acetyl-CoA/CoA ratios, 75, 76
Acetylhydrolipoate, 76
Acidic proteins, muscle Ca^{2+} binding by, 117
Actin, 117, 125, 126, 130, 131
Actin–myosin interaction, 115, 118
Action potentials, 4, 100
Acto-heavy meromyosin, 122
Actomyosin, 115, 122, 128, 130
Actomyosin adenosine triphosphatase, 118
Adenohypophysial cells, exocytosis in, 2, 20
Adenosine 3′:5′-cyclic monophosphate (cyclic AMP)
  calcium ion effects on, 100
  phosphorylase kinase phosphorylation and, 51, 56–63
  pyruvate dehydrogenase activity and, 77
Adenosine 3′:5′-cyclic monophosphate-dependent protein kinase,
  glycogen metabolism *in vivo*, role in, 69
  phosphorylase kinase activation by, 56–63
Adenosine diphosphate (ADP),
  phosphorylation, Ca^{2+} uptake and, 96
  pyruvate dehydrogenase and, 77, 83
Adenosine triphosphate (ATP),
  calcium uptake and, 89
  resting muscle, concentration in, 116
Adenosine triphosphatase (ATPase),
  calcium-activated, structure of, 111–113
  calcium movements in muscle and, 29, 38, 43, 44, 45, 46, 48
  contractile protein function of, calcium and, 115, 132
  secretion, and contractile hypothesis of, 23–24
Adenylate cyclase, 56, 69
ADP, *see* Adenosine diphosphate.
Adrenal medulla chromaffin cells, *see* Chromaffin cells.
Adrenaline,
  phosphorylase kinase control by, 56, 69
  pyruvate dehydrogenase control by, 77
Adrenergic neurones, exocytosis in, 12–13
Adrenocorticotrophic hormone, 77
Aequipecten irradians, 127
Aequorea forskalea, 37
Aequorin, as intracellular calcium indicator, 29, 33, 35, 37–42
Antagonistic cations, to Ca^{2+}, 133
Anterior pituitary, exocytosis in, 3, 16
Antigens, as exocytosis stimulants, 4
Anti-inflammatory response, 104
Antimycin A, 90, 91, 97, 99
Aragonite, 136
ATP, *see* Adenosine triphosphate.
ATP/ADP ratios, 77
ATPase, *see* Adenosine triphosphatase.
Autacoids, as exocytosis stimulants, 4

B

Bacterial nuclease, Ca^{2+}-binding by, 134
Bacterial toxins, as exocytosis stimulants, 4
Balanus nubilus (Pacific barnacle), muscle fibres, 30, 32, 35, 38–42, 44, 45, 47
Barium ion,
  calcium-activated exocytosis and, 6, 16, 23
  calcium-binding protein of muscle, and, 122
  pyruvate dehydrogenase phosphate phosphatase, and, 83
Beta-cells, of pancreas, *see under* Pancreatic.
Binding constants, for biological cations, 133
Binding strength, of Ca^{2+}, 134–136
Biosynthesis, Ca^{2+} effects on, 100
Blowfly, flight-muscle mitochondria, 92, 93
Bone formation,
  calcium binding in, 136
  mitochondrial Ca^{2+} in, 105
Butacaine, 93

C

^{45}Ca, *see* Radioactive calcium.
Cadmium ion, reaction with calcium-binding protein, 121
Caffeine, muscle fibre responses to, 29, 31, 32, 33, 34
Calcifications, mitochondrial Ca^{2+} transport and, 106–107

Calcite, 136
Calcium (ions),
   actomyosin and, 115
   aequorin as intracellular indicator in muscle, 29, 33, 35, 37–42
   affinity constants for, 115, 116, 120
   anti-inflammatory response, in, 104
   binding of, *see also* Calcium-binding protein, 83, 89, 90–93, 94, 95, 101, 116, 117, 133, 134–136
   exocytosis and exocytosis–vesiculation, in, 1–28
   hydration of, 136
   influx–exocytosis hypothesis, and, 4–7
   intracellular, mobilization in exocytosis, 17–18
   ligands and their functions, 133–138
   magnesium-activated adenosine triphosphatase and, 117–119, 122, 126–131
   mitochondrial uptake in regulation of cell function, 89–109
   movements in relation to contraction, 29–50
   murexide as indicator of, 36, 37, 97
   muscle contractile protein function, and, 115–132
   muscle contraction, movements in relation to, 29–50
   optical measurements of, 36–37
   phosphorylase kinase activation by, 51, 52–53
   pyruvate dehydrogenase regulation, in, 75–88
   radioactive, movements in muscle contraction, 29, 30–36
   sarcoplasmic, changes in contraction, 47–49
   sequestration of, 116, 134
   transport, by mitochondria, 89–90, 92, 93
Calcium-activated adenosine triphosphatase, of sarcoplasmic reticulum, 111–113
Calcium-activated exocytosis, 1–20
   chromaffin cell secretion, and, 4–7, 15
   'compound' and 'simple' forms of, 19–20
   development of concept of, 1–4
   general applicability of the concept of, 7–8
   mast cells, in, 16–20
   neurosecretory fibres and neurons, in, 8–15
   potassium ions and, 5, 6, 13–14, 16–17
   serial fusion of granules in, 19–20

Calcium-activated exocytosis—
      continued
   sources of calcium utilized in, 17–18
   variations on the theme of, 16–20
Calcium-binding protein,
   localization of, 124–126
   mitochondrial, 94–95
   troponin complex, in, 115, 116, 118–124
Calcium carrier, of mitochondrial membrane, 93–94
Calcium 1,3-diphosphorylimidazole, 135
Calsequestrin, 115, 117
Carbonate, Ca^{2+} binding to, 136
Carbon tetrachloride intoxication, 106
Carbonyl cyanide *p*-trifluoromethoxyphenylhydrazone, 97
Carboxylates, Ca^{2+}-binding strength of, 134
Cardiac surgery, mitochondrial calcium and, 106
Cartilage, mitochondrial calcium in, 104
Catecholamine secretion, calcium and, 2, 3, 13, 100
Caveolae, 10
Cell–cell conglomerates, calcium and, 134
Cell contact, calcium and, 100
Cell function, mitochondrial Ca^{2+} uptake and regulation of, 89–109
Charge/size ratio, binding strength and, 135
Chelex 100 resin, 81, 84, 95
Chemical transmitters, *see also under specific names*, calcium-dependent exocytosis stimulants, as, 4
Cholecystokinin–pancreozymin, 18
Chromaffin cells, calcium-activated exocytosis in, 2–7, 13–17
Chromogranin, 3, 13, 23
Cilia, calcium activation of, 100
Circular dichroism, of adenosine triphosphatase, 112
Cobalt ion, 83
Colchicine, effects on microtubules and secretion, 24
'Compound' exocytosis, 19–20, 21
Conconavalin, calcium binding by, 134
Condrocytes, 104
Contractile proteins, calcium ions and function in muscle of, 115–132
Contractile systems, calcium ion regulation of, 89–109
Contraction,
   calcium movements in relation to, 29–50

SUBJECT INDEX

Contraction—*continued*
  calcium movements in relation to—
                        continued
    aequorin as indicator of, 37–42
    kinetic model for, 42–46
    mechanism of, 47–49
    optical measurements of, 36–37
    radioactive calcium in study of, 30–36
  secretion and, parallels between, 23
Contraction–relaxation cycle, 101
Co-operative interactions, calcium salts in, 136
Co-ordination sphere, of calcium ion, 133, 136
Copper ion, 83
Coronary occlusion, mitochondrial calcium in, 106
Creatine phosphokinase, 103
Cross-linking, Ca^{2+} role in, 121, 122, 133, 136, 137
Cyanide, 90
Cyclic AMP, *see* Adenosine 3′:5′-cyclic monophosphate.
Cytochalasin, effects on secretion, 24–25
Cytoplasmic streaming, Ca^{2+} and, 100

D

Debranching enzyme, in muscle glycogenolysis, 51
Deoxycholate, 111, 112
Dichloroacetate, as inhibitor, 77
Dihydrolipoate, 76
Dimethylsuberimidate, 121
Dinitrophenylglycine, 120
Dinonylphthalate, 86, 87
Dipolar assemblies, 136
Dopamine β-hydroxylase, 12
Drugs, *see also under specific names*, membrane-bound Ca^{2+} and, 138

E

Earthworm, calciferous gland of, 104
Egg-shell gland, 104
EGTA, *see* Ethanedioxybis(ethylamine)-tetra-acetic acid.
Electron microscopy,
  adenosine triphosphatase, of, 111–112
  exocytotic images obtained by, 2, 3, 8–9, 10, 11, 19
  mitochondrial calcium, of, 104
  phosphorylase kinase, of, 71
'Emiocytosis', *see also* Exocytosis, 2
Endocrine glands, *see also under specific names*, calcium-dependent secretion by, 2–3, 8

End-product inhibition, of pyruvate dehydrogenase, 75
Energy coupling, mitochondrial Ca^{2+} maintenance by, 98
Energy-linked Ca^{2+} translocation, 92, 93
Enzyme regulation, by Ca^{2+}, 99–100, 103
Ethanedioxybis(ethylamine)tetra-acetic acid (EGTA),
  muscle fibre calcium efflux and, 31, 34, 35, 36, 40, 41
  muscle proteins and Ca^{2+} studies with, 117, 119, 120, 122, 123, 131
  phosphorylase kinase inactivation by, 53, 57
  pyruvate dehydrogenase and Ca^{2+} studies with, 75, 78–83, 86, 87, 88
Exchangeable calcium, in fat-cell mitochondria, 86–87, 88
Exocrine glands, calcium-dependent secretion by, 3, 4, 8, 100
Exocytosis, *see also* Calcium-activated exocytosis,
  calcium involvement in, 1–28, 136
  'compound' and 'simple', 19–20
  development of the concept of, 1–4
  dynamics of, microcinematography and, 20–21
  electron microscopical evidence of, 2, 3, 8–9, 10, 11, 19
  functional aspects of, 8–13
  microvesicle formation and, 1, 8–13
Exocytosis–vesiculation sequence, Ca^{2+} involvement in, 1–28
Exocytotic pits, 10, 11, 12, 21

F

Fat-cell mitochondria, 75
  exchangeable calcium in, 86–87
  pyruvate dehydrogenase in, 77
Ferritin-labelled antibody, 124
Ferrous ion, 83
Fish muscle, Ca^{2+} activation of, 100
α-Fluorocortisol intoxication, 106, 107

G

Glucose 1-phosphate, 51
Glutamate dehydrogenase, 84, 86, 87
Glutamic acid, in Ca^{2+}-binding protein, 95
Glutathione, 81, 86
α-Glycerophosphate dehydrogenase, 93, 100, 103
Glycogen, control of breakdown in muscle, 51–73
Glycogen particles, phosphorylase kinase in, 53, 60–61

Glycogen synthetase, 53, 59, 69, 70
Glycogen synthetase phosphatase, 63, 70
Glycogenolysis, phosphorylase kinase control of, 51–73
Glycolysis, Ca^{2+} regulation in, 103
Glycoproteins, mitochondrial Ca^{2+} transport and, 94–96
Golden hampster, evidence for exocytosis in tissues of, 9
Golgi apparatus, 11
Gramicidin, mitochondrial Ca^{2+} and, 97
GTP, *see* Guanosine triphosphate.
Guanosine triphosphate (GTP), 83

H

Heart, mitochondria in contraction and relaxation of, 101–103
High-affinity Ca^{2+}-binding sites, 85, 88, 89, 90–93, 95, 98, 116
Histamine, calcium-dependent release of, 16, 17, 18
Hormonal control,
 muscle glycogenolysis, of, 51–73
 phosphorylase kinase, of, 56
Hormones, *see also under specific names*, as exocytosis stimulants, 4
Horseradish peroxidase, 11, 12
Hydration, of Ca^{2+} ion, 133, 136
Hydrogen-bonding, in Ca^{2+} salts, 135
Hydroxyapatite, mitochondrial, 105
Hydroxyethylthiamin pyrophosphate, 76

I

I-filament, of myofibrils, 115, 124, 125, 126, 127, 128–129
Inhibitory protein, of muscle, 121, 122, 124, 125, 127
Inner mitochondrial membrane, 93, 95, 103
Insulin, 77, 85–86, 88, 100
[^{3}H]Inulin, as an extracellular marker, 86, 87
[^{125}I]Iodine, location of lipoprotein fragments with, 112
Iodoform intoxication, 106
Ionophores, 88, 95, 97, 134
Isoproterol intoxication, 106, 107

K

Kidney mitochondria, Ca^{2+} in, 106

L

Lactoperoxidase, 111, 112
Lanthanum ion, 83, 92, 95, 135
Ligands, of calcium ions, 133–138

Light chain myosin, 115, 117, 127, 129, 130, 131
Lipases, calcium activation of, 100
Lipid biosynthesis, pyruvate dehydrogenase and, 75
Lipoate, 76
Lipoprotein, from adenosine triphosphatase, 111–113
Liver mitochondria, 77, 106
Local anaesthetics, and Ca^{2+} binding, 91, 92
Lysosomes, exocytosis and, 10, 12
Lysozyme, 134

M

Magnesium ATP, 115, 118, 128
Magnesium ions,
 calcium influx–exocytosis and, 5, 14
 cell concentration of, 116, 134, 135
 deficiency of, 106, 107
 muscle adenosine triphosphatase stimulation by, 115, 116, 117, 119, 121, 122, 126, 128, 129, 131
 pyruvate dehydrogenase regulation, and, 75, 78–83, 87
 pyruvate kinase regulation by, 103
Magnesium salts, structure of, 136
Maia squinado (spider crab), muscle fibre contraction, 30, 31, 33, 34
Manganese ion, 83, 92, 94, 121, 122
Mannitol, in mitochondrial medium, 97, 99, 102
Mast cells, calcium-dependent exocytosis in, 4, 16–21
'Matrix loading', of mitochondria with Ca^{2+}, 89–90, 104–105
'Membrane flow', in exocytosis–vesiculation sequence, 1–2, 4, 10–12, 19–20, 21–23
Membrane vesiculation and exocytosis, Ca^{2+} involvement in, 21–23
Membranes, *see also under more specific names*,
 calcium and magnesium ions in, 98, 100, 137
 calcium loading in, 89–90
 phospholipids and Ca^{2+} binding in, 91
 prostaglandins and, 104
Microcinematography, of exocytosis in mast cells, 20–21
Microfilaments, Ca^{2+} activation of, 100
Micropinocytosis, 2, 3, 10, 11
Microtubules, in secretion hypothesis, 24–25, 100
Microvesicles (*see also* Synaptic vesicles), formation of, 1, 10–11

SUBJECT INDEX

Mitochondria,
  calcium transport in, 33, 89–109
  enzyme regulation in, 77–78, 103
  heart muscle contraction, in, 101–103
  liver, 77, 106
  pyruvate dehydrogenase of, 77–78
  Ruthenium Red effect on, 85
Mitochondrial calcium ions,
  accumulation of, 96–98
  affinity for, 102
  binding proteins for, 94–95
  binding sites for, 90–93, 101
  enzyme stimulation and, 103
  necrotic deposits of, 106
  pump for, 101
  pyruvate dehydrogenase phosphate activity and, 88
  pyruvate kinase control by, 103
  swelling and, 104
  transport of, 89–90, 96–99
  uptake of, 89–109
Mitochondrial matrix, Ca^{2+} from, 98, 103
Mitochondrial membrane, Ca^{2+} carrier of, 93–94
Molluscs, troponin complex of, 115, 127
Mucopolysaccharides, Ruthenium Red specificity for, 90
Multi-dentate ligands, 135, 136
Murexide, as calcium indicator, 36, 37, 97
Muscarinic drugs, 5
Muscle,
  calcium ions and contractile protein function, 115–132
  calcium movements and contraction of, 29–50
  phosphorylase kinase and glycogenolysis in, 51–73
Muscle proteins, Ca^{2+} and, 115–132
Myofibril,
  actomyosin of, 115
  Ca^{2+} regulatory role in, 101, 117–119, 128–131
  proteins of, 115, 130, 131
  sarcoplasmic reticulum and functions of, 101
Myosin,
  Ca^{2+}-dependent adenosine triphosphatase and, 118
  heavy chains of, 128
  light chains of, 115, 117, 127, 129, 130, 131

N

NADH/NAD$^+$ ratios, pyruvate dehydrogenase control by, 75, 76

Nerve transmission, Ca^{2+} effects in, 100, 101
Neurons, calcium-dependent exocytosis in, 1, 4, 8–15
Neurophysin, 8, 9, 23
Neurosecretory fibres, 1, 4, 8–15
Neurosecretory granules, 10–11
Nicotinic drugs, 5
Noradrenaline, calcium-dependent effect of, 3
Nuclease, Ca^{2+} binding by, 134

O

Oligomycin, 85, 89, 95
Ouabain, 5, 77
Oxalate, cation binding to, 136
Oxyanions, Ca^{2+}-binding by, 134

P

Pacific barnacle, see *Balanus nubilis*
Palmitoyl-L-carnitine, 77
Pancreatic acinar cells, exocytosis in, 2
Pancreatic β-cells, calcium-activated exocytosis in, 2, 3, 16
Papain intoxication, 106, 107
Parathyroid hormone, 104, 106, 107
Parvalbumin, 116, 117, 130
Pathological calcification, 106–107
Perikarya, 10, 11
Phosphate,
  Ca^{2+}-binding to, 134, 136
  pyruvate dehydrogenase regulation and, 77
  transport of, 137
Phospholipases, Ca^{2+} activation of, 100
Phospholipid, Ca^{2+} and, 95, 100, 103, 137
Phosphorylase, activation by kinase in muscle, 51–73, 100
Phosphorylase kinase,
  activation by protein kinase, 51, 57–58, 58–60
  alpha-subunit phosphorylation, 51, 58–63
  beta-subunit phosphorylation, 51, 58, 59, 60, 62–63, 71
  Ca^{2+} activation of, 51, 52–53, 68, 115
  catalysis of its own activation, 51, 56, 57, 63–64
  chromatography on DEAE-cellulose, 54
  cyclic AMP-dependent activation of, 51, 56–63, 69, 71
  deficiency of, glycogenolysis in, 50
  electron microscopy of, 71
  ethanedioxybis(ethylamine)tetraacetic acid inactivation of, 53, 57

Phosphorylase kinase—*continued*
  gamma-subunit of, 51, 54, 71
  glycogenolysis, and, 51–73
  hormonal control of, 56
  isolation of, 53–54
  molecular weight of, 51, 55–56
  muscle glycogenolysis control, in, 51–73
  myosin phosphorylation by, 66, 130
  phosphorylase interaction with, 71
  protein kinase in activation of, 51, 56–63, 69, 70
  proteolytic activation of, 51, 52, 64–65
  regulation *in vivo*, 66–68
  specificity of, 65–66
  subunit structure of, 51, 54–56
  subunit phosphorylation of, 58–63
  N-terminal sequences of, 65
Phosphorylase kinase phosphatase, 51, 61–63
Phosphorylase phosphatase, 53, 69, 70
Phosphorylative efficiency, mitochondrial Ca^{2+} and, 96
Phosphorylethanolamine, 103
Pig heart pyruvate dehydrogenase phosphate phosphatase, 75, 78–83, 87
Pinocytosis, 1, 11
Placopecten magellanicus, 127
Plasma membranes, 86, 107
Plasmocid intoxication, 106, 107
Platelets, 100, 130
Polyamine 48-80, secretion stimulant, 17–21, 25
Posterior pituitary hormones, exocytotic release of, 9, 10–11, 15
Potassium ions,
  calcium-activated exocytosis and, 5, 6, 13–14, 16–17
  muscle fibre response to, 29, 33, 34, 116
Progesterone, 100
Prolactin secretion, 17
Prostaglandins, 100, 104, 105
Protein kinase, in phosphorylase kinase activation, 51, 56–58, 69, 70, 130
Protein phosphatase, 69, 70
Proteins, *see under specific names*.
Proton movements, mitochondrial Ca^{2+} uptake and, 97
Pseudopod formation, 100
Pyrophosphate, pyruvate dehydrogenase kinase inhibition by, 77
Pyruvate decarboxylase, 76
Pyruvate dehydrogenase,
  activation of, 75
  biosynthesis and, 75
  Ca^{2+} and regulation of, 75–88

Pyruvate dehydrogenase—*continued*
  enzymes of, 76–77
  ethanedioxybis(ethylamine)tetraacetic acid, effect on 75, 78–83, 84 86, 87, 88
  extramitochondrial Ca^{2+} and, 84–86
  kinase acting on, effectors of, 77–78
  mitochondrial location of, 77–78
  molecular weight of, 76
  phosphatase acting on phosphorylated form of, 78–84
  pyruvate activation of, 84–87
  Ruthenium Red effects on, 85
Pyruvate dehydrogenase kinase, 75, 76
  effectors of, 77–78
  inhibition of, 77, 87
  mitochondrial, 78
Pyruvate dehydrogenase phosphate phosphatase,
  assay of, 78
  Ca^{2+} activation of, 75–76, 87, 103
  Ca^{2+} concentration and activity of, 78–83
  cyclic nucleotide effect on, 83
  insulin activation of, 86
  regulation *in vivo* of, 83–84
Pyruvate kinase, Ca^{2+} regulation of, 100, 103

R

Rabbit psoas muscle, 124, 125
Radioactive calcium, use of, 29, 30–36, 81, 85, 86, 88, 91, 95, 96, 99, 102
Rat epididymal adipose tissue,
  mitochondria, 84
  pyruvate dehydrogenase regulation in, 75, 77, 87
  Ruthenium Red effect on, 85
Rat heart,
  mitochondria, 102
  pyruvate dehydrogenase regulation in, 75, 77, 83
Reconstituted vesicles, adenosine triphosphatase of, 111
Renal failure, mitochondrial Ca^{2+} in, 106
Respiratory chain, Ca^{2+} and, 89, 96, 103
Respiratory inhibitors, *see also under specific names*, 89, 97
'Reverse micropinocytosis', 2, 3
'Reverse pinocytosis', 2
Rotenone, 90, 91, 97, 99
Ruthenium Red, use of, 85–86, 87, 90, 92, 94, 95, 97, 98, 103

S

Salivary glands, exocytosis in, 17–18, 20
Sarcoplasmic reticulum,

SUBJECT INDEX

Sarcoplasmic reticulum—*continued*
    adenosine triphosphatase, structure of calcium-activated enzyme, 111–113
    calcium uptake and release by, 29, 30–36, 47–49, 53, 100, 101, 102, 115, 116, 134
Sarcoplasmic vesicles, 101, 111, 112
Scallop, muscle proteins, 127, 128
Scatchard plots, 91
Secretagogues, *see also under specific names*, 2, 5
Secretion,
    calcium and contractile hypothesis of, 23–25
    calcium-activated exocytosis mechanism of, 1–28
Secretory granules,
    adenosine triphosphatase and mechanism of secretion, 23–24
    immobility of, 20–21
    membrane vesiculation and, 21–23
    serial fusion in exocytosis, 19–20
Secretory stimulants, *see also under specific names*, 2
Serial fusion of granules, in exocytosis, 19, 20, 21
'Simple' exocytosis, 19–20
Sodium ATP, 102
Sodium ion, mitochondrial Ca^{2+} release by, 99, 103
Spider crab, *see Maia squinado.*
Steroids, Ca^{2+} release of, 100
Stimulus–secretion coupling,
    calcium involvement in, 1–28
    chromaffin cells, in, 4–7, 13–14
    excitation–contraction coupling and, 4, 23
    neurosecretory fibres and neurons in, 1, 4, 8–15
Strontium ion, 78, 83, 92, 94, 121, 122
Succinate oxidase, Ca^{2+} stimulation of, 100, 103
Sugar carboxylates binding Ca^{2+}, 134
Sulphate binding of Ca^{2+}, 134
Synaptic vesicles, *see also* Microvesicles,
    adenosine triphosphatase of, 24
    exocytosis and formation of, 8–13

T

Tetanus, mitochondrial Ca^{2+} in, 106
Tetrodotoxin, 14
Thermolysin, 134
Thioacetamide intoxication, 106
Thiamin pyrophosphate, in pyruvate dehydrogenase reactions, 76, 77
Thyroxine, Ca^{2+} release of, 100
Toad, Ca^{2+} matrix loading of mitochondria of, 104
Tricarboxylic acid cycle control, 75, 103
Tropomyosin, 115, 118, 119, 123, 125, 126, 127, 128, 130, 131
Troponin and troponin complex, 30, 66, 101, 102, 115–119, 123–127, 129–131
Trypsin, 112
Tumour calcinosis, 106
Tumour mitochondria, Ca^{2+} in, 106

U

Uncouplers, 90, 95, 97, 98
Uranium intoxication, 106

V

Valinomycin, 97
Vasopressin, Ca^{2+} release of, 100
Vesiculation, *see also* Exocytosis–vesiculation, 10–12, 21–23
Vinoblastine, 25
Vitamin D, mitochondrial Ca^{2+} and, 106, 107

X

X-ray diffraction, Ca^{2+}-binding protein, 124

Y

Yeast mitochondria, Ca^{2+}-binding sites of, 92, 93

Z

Zinc ion, 83, 122